Communication Techniques in Farm Extension

Dr. T. Rathakrishnan
Dr. M. Israel Thomas
Dr. L. Nirmala

SCIENTIFIC PUBLISHERS

Published by:

SCIENTIFIC PUBLISHERS (INDIA)
5A, New Pali Road, P.O. Box 91
Jodhpur 342 001, India
E-mail: info@scientificpub.com
Website: www.scientificpub.com

Print: 2020

ISBN: 978-81-72336-07-3
eISBN: 978-93-86347-48-0

Printed in India

FOREWARD

The goal of socio – economic development can be successfully achieved only if people extend support to the efforts taken by governments. Participation of the citizen is highly inevitable in fulfilling the envisaged developmental tasks. Developmental schemes fail to a certain extent due to the dearth of people's participation and co-operation. Here in lies the need for effective communication between policy makers and stakeholders. The mass media existing in our country viz., newspaper, Radio, TV, and film can play a vital role in accelerating national development. Development Journalism, Cyber Extension and problem tree analysis is meant to foster the holistic development of human potential vis-à-vis the economic prosperity of the country.

Development Journalism focuses on the needs of the poor, the deprived and marginalized and emphasizes their effective participation in developmental planning. Cyber extension emphasizes on all the recent developments in communication technology with respect to extension. The problem tree analysis is ready reference for field level functionaries for solving field level problems. The team headed by Dr. T. Rathakrishnan. Prof. & Head, Dept. of Agricultural Extension & Rural Sociology, Agricultural College and Research Institute, Madurai – 625 104 has been taken a efforts to bring together the aspects of development journalism and Cyber Extension and problem tree analysis under one umbrella by way of compiling and publishing in the book form. This book systematically presents the recent developments in the area of journalism, Cyber Extension and problem tree analysis.

I hope this publication will serve as a valuable reference material on Developmental journalism, Cyber Extension and problem tree analysis not only for scientists, research workers but also to student community, NGO's and policy makers.

I congratulate the authors for their valuable efforts to publish this book.

Dr. S. Jebaraj
Director
Tamil Nadu Rice Research Institute (TNAU),
Aduthurai - 612 101

EXPERT OPINION ON THE BOOK ENTITLED
"COMMUNICATION TECHNIQUES IN FARM EXTENSION"

A text book need to be done atleast three things: first, to present role of the body of data and interpretations of the same data to reasonably satisfying the critical demands of professional colleagues; second, to form the frame work of theory which students can use in interpreting data after the course is completed; third, it should present data, interpretation, and theory in a manner that stimulates the challenges of the students.

Agricultural Extension is one of the largest non formal concepts of solving extended educational systems in the world. It is generally concerned with transferring knowledge and research to farmers but may also include services to other target agencies such as farm families and rural youths, as well as serve for developing rural community resources.

One of the basic management functions of good extension system is to improve the quality of extension contact, which in turn means establishing repeated contacts with the clientele, building confidence, communicating improved technologies and providing motivation for its adoption. To do this, anyone must be fully understand the process of communication at the problem solving level and have the knowledge in communication techniques with regard to farm extension. This book satisfies the above requirements.

This book comprehensively covers three broad areas which one could not find from any single source. The major areas includes Farm Journalism which deals with basic concept and dimensions, principles, genesis and growth, journalism and mass media, ideals, functions and canons of journalism, ethics of journalism, art of speech, readability meaning, definition and formula for readability, newspaper makeup, E-journalism - online journalism, types of leads, writing feature stories, news, types of news, photo journalism, reporting - qualities of a good reporter, reporting skills, editing, balancing news paper. The cyber extension, information and communication technology (ICTs) covers genesis with development of instructional technologies, application of multimedia in agricultural extensions, Interactive video disc, cyber extension, information and communication technologies (ICTs), expert system, agri-portal, tele conferencing and its application, audio conferencing, video conferencing, trends of development in ICT. Finally, the new topic which is not covered in any of the text book of this nature is transfer of technology efforts - problems tree analysis. It will be of immense helpful to the field level extension functionaries and students communities to strengthen their communication skills.

With rapid strides in agricultural sciences virtually, there is an information explosion. Mass media including newspaper, magazine, radio, TV are coming out with more and more agricultural information support programmes. This book act as an interface between media managers and professionals for the development of agricultural front.

This book is of more useful to the students of undergraduate, post graduates and research scholars in the field of extension, cyber extension and field problem solving. Also, this book would prove to be helpful to the extension workers in planning and streamlining communication strategies. Keeping in view, the fast developing nature of communication technologies, recent advancements in the field of communication technology are also given much emphasis in the present book.

Advances in Communication Techniques in Farm Extension is undoubtedly a learner friendly book. In order to help the learner to cope up with the wide range of concepts and skills that covers all paradigm shift of agricultural communication, this book has been brought out.

In addition to the students' community, this book may satisfy the requirements of all those organizations and individuals interested to get acquainted with the communication techniques to promote extension services leading to considerable impact in the research – extension – client systems.

I warmly recommend the book entitled "**Advances in Communication Techniques in Farm Extension**" to the students and professionals of all shades everywhere.

Dr. G. SELVARAJ
Director
Open and Distance Learning
Tamil Nadu Agricultural University, Coimbatore

PREFACE

In many Asian countries, there is now a more rapid economic growth than there has ever been in the so called developed countries. This results in major changes in agriculture and therefore agricultural extension also has no change. During the last few decades means of communication have really been revolutionized and the process still continues. Our communication media have never assumed the crucial responsibility of defining, reviewing its role with regard to social concerns and peoples problems. Readers may not look to farm magazines for literacy writing still it should be entertaining as well as informative. Taking stock of past experiences the present situation, future needs of the country and aspirations of the farmers at large , it is essential that the extension activities are undertaken by various organizations that can render efficient services to farming community.

Computers can be used to assist agricultural decision making through tools and techniques as optimization, simulation and computer aided drafting etc., research in agriculture enquires some uses of computers for crop modeling, GIS, MIS, etc., now a days media management is effectively carried out through IT applications. Audio, Video medium for technology transfer will bring more credibility on the part of technology scientists and the medium. Case study method can be used which can know precisely the cause and effect relationship of a particular phenomenon. It is good method of collecting information about the process or individuals or groups. In the same way extension literature is the major form of communication method. The major objectives of printed literature is to communicate the ideas in such a manner to attract the attention of the farmers, help them to understand and remember the message and ultimately facilitate them to take appropriate decision.

How to write a new story suggests the characteristics of the news, principles of writing, different types of lead, elements of style and also readability measurement. Agricultural journalism provides agricultural related information to audience such as consumers, legislators, agri-business, community groups etc. Today a large number of accredited journalists together with countless stingers, communicators and other seek out report and publish the happenings.

With the advent of many facets of IT it is necessary that planners and policy makers should develop a crucial a critical perspectives and understanding about the potential and challenges of information system and technologies so that they can understand anticipate and push the technology process. The process of technology transmission is enhanced greatly by the voice, data and video capabilities made available by satellite and micro wave transmission., knowledge expansion, information technology and information dissemination are the important dimensions of development system. Magazine journalism not much different form news paper journalism stingers and freelancer works by many agencies. With out sourcing and contacting becoming common, free lancers known as independent professionals are experiencing a welcoming change. Development reporting has tasks to perform, inform, interpret and promote. It is otherwise called advocacy journalism. Through the government has spent billions of money on different development program not much has been achieved in terms of making master literature and uplifting their living standards.

The book Communication Techniques in Farm Extension has been created to provide access to information about contemporary topic in Farm Journalism, Cyber Extension and Transfer of Technology to practioners and students at all levels in Journalism and ICTs and IT have a need to know what is happening today, in addition to historical treatments within the literature.

With rapid strides in agricultural sciences virtually there is an information explosion; mass media including news paper, magazine, radio, TV are coming out with more and more agricultural information support programmes. This book act as an interface between media mangers and professionals for the development of farm front.

AUTHORS

CONTENTS

I. FARM JOURNALISM

Chapter - 1

Journalism Concepts and Principles

1.1. Definition and Meaning

The world today is in the midst of information explosion. As recently as world war II most people relied almost exclusively on newspapers for information relating to current events. Today the dimensions of communication have expanded to include radio, TV, news magazines, cable vision, news letters, public relations, some forms of advertising, microfilm applications and spectacular developments in photography.

Journalism is the key to communication. Good journalism techniques are the same for TV, radio and newspaper whether large or small, daily or weekly. All the systems of communication no matter where the news develops have the same basic methods of gathering and organising the facts. They also share a common standard of ethical practice and journalistic rights and duties.

Journalism is a fascinating field that takes its practitioners to the places where things are happening and to the people who are making the history (Shamsul Hasan, 1990).

Journalism encompasses fields ranging from reporting with words and photographs to editing and from news papers to TV (Spencer Cremp, 1914).

Journalism is the report of the things as they appear at the moment of writing and not a definitive study of situation (Warren *et al.,* 1979).

Journalism is the eyes, ears and curiosity of the public and must be so broad in their outlook that they can translate events in many fields.

Etymologically journalism may be described as a compendium of information selected to entertain a target audience with a wide range of subject interests and presented in ways which may be appealing to different sections of the target audience (Sudarsanam, 1990).

According to Veerabadriah (1990) Journalism is the "systamatic" and "reliable", gathering, writing, interpreting, processing and dissemination of public information, public opinion and public entertainment for publication, in newspapers, magazines, broadcasts and telecasts.

The purpose of journalism is to inform, to guide, and to assist business through advertising.

It is a discipline of collecting, analyzing, verifying, and presenting news regarding current events, trends, issues and people.

Journalism has as its main activity the reporting of events — stating who, what, when, where, why and how, and explaining the significance and effect of events or trends. Journalism exists in a number of media: newspapers, television, radio, magazines and, most recently, the World Wide Web through the Internet.

The subject matter of journalism can be anything and everything, and journalists report and write on a wide variety of subjects: politics on the international, national, provincial and local levels, economics and business on the same four levels, health and medicine, education, sports, hobbies and recreation, lifestyles, clothing, food, pets etc;

As for the future of journalism, computers are increasingly being used to support this discipline. For example, Artificial Intelligence (AI) is used to support editorial operations, such as "Artificial Intelligence Journalist" (AIJ), a computer software developed by publisher NewsRx.

1.2. Major Dimensions of Journalism

Journalism has many facets and encompasses many areas

1. News/Feature/Magazine/editorial writing
2. Broadcast journalism
3. Language journalism
4. Photo journalism
5. Sports journalism
6. Development journalism
7. Technical journalism

(i) Farm journalism

(ii) Business/trade journalism

(iii) Scientific journalism

(iv) Home economics journalism

(v) Engineering journalism

The specialised aspects of journalism are usually referred to as the 'Technical journalism Field' and it deals with technical materials in agriculture, Home economics, engineering, Science, industry, trade, business and the professions.

Going by the definition of most dictionaries, Journalism would be confined to the writing for newspapers and magazines and so, it was until the 1920's. But today the average person interprets journalism to include not only newspapers and magazines, but also radio, TV, Cassette, public relations, advertising, Book publishing and education directed to these fields. Newspapers however offer widest range in writing experiences, and more journalists start in this field and transfer to radio, TV, advertising Public Relation (PR), Newspaper methods or techniques form the basic training for many fields of journalism.

The details about the major dimensions are given below

Print journalism

Print journalism can be split into several categories: newspapers, news magazines, general interest magazines, trade magazines, hobby magazines, newsletters, private publications, online news pages and others. Each genre can have its own requirements for researching and writing reports.

For example, newspaper journalists in the United States have traditionally written reports using the inverted pyramid style, although this style is used more for straight or hard news reports rather than features. Written hard news reports are expected to be spare in the use of words, and to list the most important information first, so that, if the story must be cut because there is not enough space for it, the least important facts will be automatically cut from the bottom. Editors usually ensure that reports are written with as few words as possible. Feature stories are usually written in a looser style that usually depends on the subject matter of the report, and in general granted more space

News magazine and general interest magazine articles are usually written in a different style, with less emphasis on the inverted pyramid. Trade publications can be more news oriented, while hobby publications can be more feature oriented.

Broadcast journalism

Radio journalists must gather facts to present them fairly and accurately, but also must find and record relevant and interesting sounds to add to their reports, both interviews with people involved in the story and background sounds that help characterize the story. Radio reporters may also write the introduction to the story read by a radio news anchor, and may also answers questions live from the anchor.

Television journalists rely on visual information to illustrate and characterize their reporting, including on-camera interviews with people involved in the story, shots of the scene where the story took place, and graphics usually produced at the station to help frame the story. Like radio reporters, television reporters also may write the introductory script that a television news anchor would read to set up their story. Both radio and television journalists usually do not have as much "space" to present information in their reports as print journalists.

On-line journalism

The fast and vast growth of the Internet and World Wide Web has spawned the newest medium for journalism, on-line journalism. The speed at which news can be disseminated on the web, and the profound penetration to anyone with a computer and web browser, have greatly increased the quantity and variety of news reports available to the average web user.

The bulk of on-line journalism has been the extension of existing print and broadcast media into the web via web versions of their primary products. New reports that were set to be released at expected times now can be published as soon as they are written and edited, increasing the deadline pressure and fear of being scooped many journalists must deal with.

The growth of blogs as a source of news and especially opinion on the news has forever changed journalism. Blogs now can create news as well as report it, and blur the dividing line between news and opinion. The debate about whether blogging is really journalism rages on.

Feature journalism

Newspapers and periodicals often contain features written by journalists, many of whom specialize in this form of in-depth journalism.

Feature articles usually are longer than straight news articles, and are combined with photographs, drawings or other "art." They may also be highlighted by typographic effects or colors.

Writing features can be more demanding than writing straight news stories, because while a journalist must apply the same amount of effort to accurately gather and report the facts of the story, the reporter must also find a creative and interesting way to write the article, especially the lead, or the first one or two paragraphs of the story. The lead must grab the reader's attention yet accurately embody the ideas of the article. Often the lead of a feature article is dictated by its subject matter. Journalists must work even harder to avoid clichéd images and words when writing the lead and the rest of the article.

In the last half of the 20th Century the line between straight news reporting and feature writing blurred as more and more journalists and publications experimented with different approaches to writing an article. Tom Wolf, Gay Talese, Hunter S. Thompson and other journalists used many different approaches for writing news articles. Urban and alternative weekly newspapers went even further blurring the distinction, and many magazines fan more features than straight news. Some television news shows experimented with alternative formats, and many TV shows that claimed to be news shows were not considered as such by many critics, because their content and methods did not adhere to accepted journalistic standards. A majority of newspapers still maintain a clear distinction between news and features, as do most television and radio news organizations.

Sports journalism

Sports journalism covers many aspects of human athletic competition, and is an integral part of most journalism products, including newspapers, magazines, and radio and television news broadcasts. While some critics don't consider sports journalism to be true journalism, the prominence of sports in Western culture has justified the attention of journalists to not just the competitive events of sports, but also to athletes and the business of sports.

Sports journalism in the United States has traditionally been written in a looser, more creative and more opinionated tone than traditional journalistic writing; the emphasis on accuracy and underlying fairness is still a part of sports journalism. An emphasis on the accurate description of statistical performances of athletes is also an important part of sports journalism.

Science journalism

Science journalism is a relatively new branch of journalism, in which journalists' reporting conveys information on science topics to the public. Science journalists must understand and interpret very detailed, technical and sometimes jargon-laden information and render it into interesting reports that are comprehensible to consumers of news media.

Scientific journalists also must choose which developments in science merit news coverage, as well as cover disputes within the scientific community with a balance of fairness to both sides but also with a devotion to the facts.

Many, but not all, journalists covering science have training in the sciences they cover, including several medical doctors who cover medicine.

Investigative journalism

Investigative journalism, in which journalists investigate and expose unethical, immoral and illegal behaviour by individuals, businesses and government agencies, can be complicated, time-consuming and expensive, requiring teams of journalists, months of research, interviews (sometimes repeated interviews) with numerous people, long-distance travel, computers to analyze public-record databases, or use of the company's legal staff to secure documents under freedom of information laws.

Because of its inherently confrontational nature, this kind of reporting is often the first to suffer from budget cutbacks or interference from outside the news department. Investigative reporting done poorly can also expose journalists and media organizations to negative reaction from subjects of investigations and the public, and accusations of gotcha journalism. When conducted correctly it can bring the attention of the public and government problems and conditions that the public deem need to be addressed, and can win awards and recognition to the journalists involved and the media outlet that did the reporting.

Gonzo journalism

Gonzo journalism is a type of journalism popularized by the American writer Hunter S. Thompson, author of Fear and Loathing in Las Vegas, *Fear and Loathing: On the Campaign Trail of '72,* among other stories and books. Gonzo journalism is characterized by its punchy style, rough language, and ostensible disregard for conventional journalistic writing forms and customs. Gonzo journalism attempts to present a multidisciplinary perspective on a particular story, drawing from popular culture, sports, political, philosophical and literary sources. Gonzo journalism has been styled eclectic or untraditional. It has a good deal in common with on-line journalism

'Celebrity' or 'People' journalism

Another area of journalism that grew in stature in the 20th Century is 'celebrity' or 'people' journalism, which focuses on the personal lives of people, primarily celebrities, including movie and stage actors, musical artists, models and photographers, other notable people in the entertainment industry, as well as people who seek attention, such as politicians, and people thrust into the attention of the public, such as people who do something newsworthy.

Celebrity journalism differs from feature writing in that it focuses on people who are either already famous or are especially attractive, and in that it often covers celebrities obssessively, to the point of these journalists behaving unethically in order to provide coverage. Paparazzi, photographers who would follow celebrities incessantly to obtain potentially embarrassing photographs, have come to characterize celebrity journalism.

Ambush journalism

Refers to aggressive tactics practiced by journalists to suddenly confront with questions people who otherwise do not wish to speak to a journalist. The practice has particularly been applied by television journalists

The practice has been sharply criticized by journalists and others as being highly unethical and sensational, while others defend it as the only way to attempt to provide those subject to it an opportunity to comment for a report.

Gotcha journalism

Refers to the deliberate manipulation of the presentation of facts in a report in order to portray a person or organization in a particular way that varies from an accurate portrayal based on balanced review of the facts available. It particular is applied to broadcast journalism, where the story, images and interviews are tailored to create a particular impression of the subject matter.

It is considered highly unethical to engage in gotcha journalism. Many subjects of reporting have claimed to have been subjected to it, and some media outlets are guilty of deliberately biased reporting.

2. JOURNALISM PRINCIPLES

A former correspondent of the Associated Press of America, Wes Gallgher, has laid down the following principle: "Journalism requires a discipline of mind. It requires a sympathetic understanding of mankind, and also a cold objective analysis of how mankind meets the problem of this age".

"Facts are sacred. comment is free". That is what the great editor of The Manchester Guardian, C.P. Scott had said. Facts must not be tampered with. News must be reported with complete objectivity, without any distortion. A journalist should not mix news and views. He must be careful enough to clearly distinguish between the two. A well-known American journalist, Leonard Silk, commenting on The New York Times, wrote: "Everybody talks about the separation of the Church and the State, and at The Times Church refers to the editorial page and State to the news department.

Maintenance of high professional standards is certainly a major responsibility of the publisher and the editor. On the one hand journalists ought to be vigilant to safeguard the freedom of the press and on the other they must see that this freedom is used and never misused, that they always act with a great sense of responsibility.

Over time journalists have developed nine core principles what might be described as the theory of journalism:

1. JOURNALISM'S FIRST OBLIGATION IS TO THE TRUTH

Democracy depends on citizens having reliable, accurate facts put in a meaningful context. Journalism does not pursue truth in an absolute or philosophical sense, but it can and must pursue it in a practical sense. This "journalistic truth" is a process that begins with the professional discipline of assembling and verifying facts. Then journalists try to convey a fair and reliable account of their meaning, valid for now, subject to further investigation. Journalists should be as transparent as possible about sources and methods so audiences can make their own assessment of the information. Even in a world of expanding voices, accuracy is the foundation upon which everything else is built context, interpretation, comment, criticism, analysis and debate. The truth, over time, emerges from this forum. As citizens encounter an ever greater flow of data, they have more need not less for identifiable sources dedicated to verifying that information and putting it in context.

2. ITS FIRST LOYALTY IS TO CITIZENS

While news organizations answer to many constituencies, including advertisers and shareholders, the journalists in those organizations must maintain allegiance to citizens and the larger public interest above any other and they have to provide the news without fear or favor. This commitment to citizens first is the basis of a news organization's credibility, the implied covenant that tells the audience the coverage is not slanted for friends or advertisers. Commitment to citizens also means journalism should present a representative picture of all constituent groups in society. Ignoring certain citizens has the effect of disenfranchising them. The theory underlying the modern news industry has been the belief that credibility builds a broad and loyal audience, and that economic success follows in turn. In that regard, the business people in a news organization also must nurture not exploit their allegiance to the audience ahead of other considerations.

3. ITS ESSENCE IS A DISCIPLINE OF VERIFICATION

Journalists rely on a professional discipline for verifying information. When the concept of objectivity originally evolved, it did not imply that journalists are free of bias. It called, rather, for a consistent method of testing information a transparent approach to evidence precisely so that personal and cultural biases would not undermine the accuracy of their work. The method is objective, not the journalist. Seeking out multiple witnesses, disclosing as much as possible about sources, or asking various sides for comment, all signal such standards. This discipline of verification is what separates journalism from other modes of communication, such as propaganda, fiction or entertainment. But the need for professional method is not always fully recognized or refined. While journalism has developed various techniques for determining facts, for instance, it has done less to develop a system for testing the reliability of journalistic interpretation.

4. ITS PRACTITIONERS MUST MAINTAIN AN INDEPENDENCE FROM THOSE THEY COVER

Independence is an underlying requirement of journalism, a cornerstone of its reliability. Independence of spirit and mind, rather than neutrality, is the principle that journalists must keep in focus. While editorialists and commentators are not neutral, the source of their credibility is still their accuracy, intellectual fairness and ability to inform not their devotion to a certain group or outcome. In our independence, however, we must avoid any tendency to stray into arrogance, elitism, isolation or nihilism.

5. IT MUST SERVE AS AN INDEPENDENT MONITOR OF POWER

Journalism has an unusual capacity to serve as watchdog over those whose power and position mostly affects citizens. The Founders recognized this to be a rampart against despotism when they ensured an independent press; courts have affirmed it; citizens rely on it. Journalists have an obligation to protect this watchdog freedom by not demeaning it in frivolous use or exploiting it for commercial gain.

6. IT MUST PROVIDE A FORUM FOR PUBLIC CRITICISM AND COMPROMISE

The news media are the common carriers of public discussion, and this responsibility forms a basis for special privileges. This discussion serves society best when it is informed by facts rather than prejudice and supposition. It also should strive to fairly represent the varied viewpoints and interests in society, and to place them in context rather than highlight only the conflicting fringes of debate. Accuracy and truthfulness require that as framers of the public discussion Journalists not neglect the points of common ground where problem solving occurs.

7. IT MUST STRIVE TO MAKE THE SIGNIFICANT INTERESTING AND RELEVANT

Journalism is storytelling with a purpose. It should do more than gather an audience or catalogue the important. For its own survival, it must balance what readers know they want with what they cannot anticipate but need. In short, it must strive to make the significant interesting and relevant. The effectiveness of a piece of journalism is measured both by how much a work engages its audience and enlightens it. This means journalists must continually ask what information has most value to citizens and in what form. While journalism should reach beyond such topics as government and public safety, a journalism overwhelmed by trivia and false significance ultimately engenders a trivial society.

8. IT MUST KEEP THE NEWS COMPREHENSIVE AND PROPORTIONAL

Keeping news in proportion and not leaving important things out are also cornerstones of truthfulness. Journalism is a form of cartography: it creates a map for citizens to navigate society. Inflating events for sensation, neglecting others, stereotyping or being disproportionately negative all make a less reliable map. The map also should include news of all our communities, not just those with attractive demographics. This is best achieved by newsrooms with a diversity of backgrounds and perspectives. The map is only an analogy; proportion and comprehensiveness are subjective, yet their elusiveness does not lessen their significance.

9. ITS PRACTITIONERS MUST BE ALLOWED TO EXERCISE THEIR PERSONAL CONSCIENCE

Every journalist must have a personal sense of ethics and responsibility - a moral compass. Journalists must be willing, if fairness and accuracy require, to voice differences with their colleagues, whether in the newsroom or the executive suite. News organizations do well to nurture this independence by encouraging individuals to speak their minds. This stimulates the intellectual diversity necessary to understand and accurately cover an increasingly diverse society. It is this diversity of minds and voices, not just numbers that matters.

Genesis and growth of Journalism in India.

Growth of English Press

To begin with, English newspapers were published by British in India to convey news from Britain and Europe to those residing here. The Christian missionaries were among the first to start newspapers in English. With a circulation of a few hundred copies per newspaper, the English press enjoyed popular support among the readers from the initial stages itself.

If the proprietors of the newspapers were those who had fallen from the grace of the officers of the East India Company, they were likely to criticize the operations of the company in the subcontinent, in their papers. To this category belongs James Augustus Hickey, who takes the credit for having started the first English newspaper in British India, in the year 1780. The first English newspaper in India was also the first newspaper in India.

Hickey brought out the Bengal Gazette on January 29, 1780. He later renamed it as the Oriental Calcutta General Advertiser. It appeared weekly, every Saturday. He attacked the East India Company vigorously through his paper. Finally, when he indulged in casting aspersions on the wife of the then Governor-General, Warren Hastings, he was forcefully deported on the Governor-General's orders. A prominent merchant by the name of James Silk Buckingham, was another pioneer of English newspapers in India. He started the Calcutta Journal as its proprietor editor. From October 2,1818 till November 9,1823 (the day it ceased publication), the paper enjoyed uncommon popularity as it carried items of local Indian interest and did not play favorites with any political party. However, on March 1,1823, Buckingham was deported after Chief Secretary to the Government, John Adams announced his regulations on the free press by way of compulsory pre-publication licensing.

Bengal Harkaru, Friend of India and Bengal Herald were other newspapers that were quite popular in the beginning of the nineteenth century. Bombay Herald was Bombay's first newspaper which was started in 1789. Madras Courier was the first newspaper to be published from Madras in 1785. The first regular venture of the natives in the field of English journalism was Reformer started by Prasunna Coomar Tagore and Bholanath Sen in 1831. Earlier, in 1829, Prasunna Coomar Tagore and Raja Ram Mohan Roy had invested in Bengal Herald along with three other Indians and one British who was officiating as the editor. But, the authorities would not entertain native association with any publication and prevented the Indians from running the Bengal Herald. The Reformer, however, was popular with the Europeans as it carried moderate views. Prior to the arrival of the Reformer, the first Indian-owned English newspaper appeared in. 1816. It was called Bengal Gazette after Hickey's venture, and was owned by Gangadhar Bhattacharya.

Contribution to India's Independence

In the nineteenth century, when just a few Indians were acquainted with the English language, some educated leaders ventured to publish English newspapers. Their efforts were twice successful, as these publications put across the majority viewpoint to the British rulers and also initiated British influenced Indians into the freedom struggle.

Soon enough the fledgling press invited trouble in the form of severe government restrictions on the freedom of information and the right to criticize. Although licensing was not required, registration of newspapers was made compulsory in 1835 by Sir Charles Metcalfe, the interim Governor-General in Place of Lord William Bentinck. Licensing was reintroduced by an Act in 1857. The Press and Registration of Books Act, 1867 was essentially directed against the Vernacular Press. An interesting reaction to the Vernacular Press Act of 1878 which was promulgated by Lord Lytton, is the case of the bilingual Amrita Bazar Patrika. It was being published in Bengali and English when it suddenly converted itself into an exclusively English daily in order to escape the harsh conditions of the Act.

Leading members of the elite in society enthusiastically contributed their views to Indian-owned English newspapers. This in turn generated tremendous support for the nationalist movement. While the government clamped restrictions on the English and Vernacular Press, those papers which were owned by Anglo-Indians were exempted since they were not too critical of the British government.

The struggle for freedom gained momentum in the last few years of the nineteenth century, and this to a large extent was one of the views which appeared in English language newspapers. M.K. Gandhi, who had regularly published Indian Opinion when he was in South Africa, continued to bring out the paper even after returning to India. In addition, he started its sister publication, Harijan. Gandhiji exhorted the press in general to support the struggle for freedom and to continue to appear even in the form of handwritten pamphlets in case the printing equipment was confiscated. The English press responded to the call to fight for a free India. During the Civil Disobedience Movement, the Press (Emergency Powers) Act, 1931 was enforced. In spite of the constraints, the English papers continued to criticize the British government fearlessly.

Several newspapers which started with the singular purpose of giving a boost to the nation's struggle for freedom have survived to this day. To name just a handful of such papers, we have the examples of centenarians like The Times of India of Bombay (1838), Amrita Bazaar Patrika of Calcutta (1868), The Statesman of Calcutta (1875), The Hindu of Madras (1878), The Tribune of Chandigarh (1881).

If Calcutta gave birth' - to Indian journalism, Bombay became the hub of social reform. Dadabhai Naoroji, Behram Malabari, M.G. Ranade and a few others began reforming society with the help of their writings and speeches. Dadabhai Naoroji's own monthly journal, the Voice *of India,* came out on February 1, 1882 aiming at securing fair hearing and justice for India.

The Hindu of Madras started in 1878 as a weekly by G. Subramaniya Iyer and his enthusiastic team supported the raising of the age for marriage from 10 to 12 year. In those days it was thought an unorthodox and rebellious view, not consistent with the ideas of the leaders of the Indian society. G. Subramaniya Iyer was in the forefront of the social reform in the south. *The Hindu,* one of the most respected and prestigious of the English, Dailies of India, still holds sway over the educated and the enlightened.

The Madras Times, started in 1860, came out again in 1868 as The Madras Mail. It became simply. The Mail in 1950, Hence. The Mail or The *Madras Mail* was the oldest of the English Newspapers from Madras. It had been serving the, cause of the European and other English speaking communities in and around Madras.

The pioneer of Allahabad, Madras Mail of Madras and the Civil and Military Gazette of Lahore were very much in the forefront, providing quality news reports. One must make mention here of Annie Besant's New India, which carried some fiery editorials in its time.

In general, all the English newspapers used forceful language and served to direct and organize the energies of the people who then moved nearer to their aim of self governance.

The Indian-owned newspapers were cheaper than the British-owned ones. Naturally, this fact benefited the former category in terms of increased sales.

The single-minded manner in which the English newspapers rallied around to free the nation from the imperialist hold is a lesson in itself. Never since, has any other goal enthused the entire English press enough to raise the profession to the height of a mission.

Growth of Tamil Press

Tamil is one of the oldest, richest and the most highly cultivated of Dravidian languages of South India. The first Tamil magazine was a monthly, Tamil Patrika, started in 1831. The progress of the Tamil press was accelerated in the last decade of the 19th century with the starting of Swadesamitran in 1882 founded by the great G. Subramanya Aiyer, who converted it into a daily in 1899. The next Tamil daily made its mark only in 1917 entitled `Desabhaktan', but was forced to be closed down in 1926; that year Tamil Nadu under Varadarajulu Naidu started as a daily. But it was Swadesamitran which had the highest circulation, thanks to its stalwart editor C.R. Srinivasan. In 1936, the Indian Express group started Dinamani, a daily, in September, 1934. A.N. Sivaraman, its editor, spared no effort to make it a leading Tami daily like Swadesamitran.

In Madurai, a Tamil daily known as Dina Thanthi was born in 1940. Within a short time simultaneous editions were brought out from Madras, Coimbatore, Salem and Tiruchirapalli. Becoming immensely popular under the editorship of T.R. Bheem Singh, it overtook the two giants Swadesamitran and Dinamani.

Among the Tamil weeklies, it is Ananda Vikatan which achieved notable success. Started in 1924, it has always been identified with its founder editor, S.S. Vasan, the movie tycoon. Kalki, started by R. Krishnamurti in 1941, is another popular

Tamil weekly. But the largest circulated Tamil weekly is Kumudum, started in 1947.

Development in Journalism

The chronology of important events gives at a glance in the field of journalism in India. Events are given media wise press, publications, advertising, radio, TV, and films and within the media, the arrangement is in chronological order.

Press

1550 The first printing press was brought to India. Printing was intr-oduced in India (1556). First book printed in India (Goa) in 1557.

1766 First modern newspaper in India was brought out. The first Indian paper 'Bengali Gazette' or 'Calcutta General Advertiser' popularly known as Ricky's Gazette appeared (1780).

1908 Newspapers (Incitement to Offences) Act.

1941 First Continuing Professional Journalism Institute in Punjab University.

1955 The Working Journalists Act. Formation of Press Guild of India.

1970 Wall Newspaper-Hamara Desh (English) and Apna Desh (Hindi)

1972 National Union of Journalists (India)

1976 'Employment News' (English) & 'Rojgar Samachar' (Hindi)

1978 Second Press Commission appointed

Publications

1557 First book printed in India (Goa) 'Doutrina Christa'

1957 National Book Trust

1972 First world Book Fair in New Delhi

Advertising

1780 First advertisement in India 'Bengal Gazette'. Advertising Agency Association of India

1945 Advertising Branch of I & B Ministry (DAVP)

1955 New Advertising policy of the .Government of India

1978

Radio The government gave license to Indian Broadcasting Company, a private concern, to open broadcasting stations in India (in 1930) the Company went into liquidation.

1926 Stated operating Indian Broadcasting Service

1930 Indian State Broadcasting Service was renamed All India Radio.

1936 Report of the Committee on Broadcasting and Information Media on Radio and Television (Chanda Committee)

1966 A 'Radio Farm Forum' - Farm

1968 First book printed in India (Goa) 'Doutrina Christa' and Home broadcasting units started

Television

1959 Television on experimental basis commenced in India

1967 Farm programme - dissemination of information on agriculture.

1975 The Satellite Instructional Television Experiment covering 2,400 villages in six States commenced (the SITE programme discontinued from August 1976).

1977 Verghese Committee appointed on Radio and Doordarshan

1978 Verghese Committee on AIR and Doordarshan - report.

Films

1912 The first Indian film 'Pundalik' shown

1918 The Cinematograph Act 1918

1931 The first Indian sound film 'Alam Ara' made

1947 Films Division

1960 Hindustan Photo Film Manufacturing Company

1966 Report of the Committee on Broadcasting & Information Media (Chanda Committee) on Documentary Films & News Reels.

1969 First Cinemascope - Kagaz-ke-phool (Guru Dutt).

1978 New guidelines on film censorship.

1990 FFI - Foundation for Freedom of Information. Non political and communal body formed on Feb 8th with a view to encourage free flour information to the public.

Chapter – 2

Journalism and Mass Media

Role of media in agriculture / rural development

UNESCO Report on Rural Development and flow of communication reported that.

- The voice of the peasant is not heard in the cities and at the seat of the Government. Populations with common background and interests are cut off from each other.
- Even the flow of communication from the centre is insufficient to meet the needs and interests of rural people who lack information on world affairs which directly affect their future and well being.
- They are deprived of learning not only their duties to the country but also their rights as citizens.

In order to communicate with the people of rural areas, communication development is very essential. Communication is very important not just to give messages, but for the valuable feed back that provides a measure of the progress made by the rural society.

In the whole communication development process rural people are many ways disadvantage although they constitute a majority of the population in the Third world. If the social and economic development plans of various countries of Asia are to produce tangible results, it should be recognised that the broadcast media have to play a vital role (both radio and TV) and hence they should be treated as part of the countries basic facilities so as to promote rural development.

Communication development in the third world countries should be aimed towards giving priorities to the field of Rural Development (RD) which comprises health, agriculture, industry,

Science, education and concentrate effort in eradicating poverty, combating illiteracy while making every possible effort for the enrichment of the native cultures, values and traditions.

Recent studies in developed countries have revealed that the media of communication particularly the modern media like newspaper, cinema, radio and TV have played supportive role in the successful involvement of the people in their developmental programs.

Daniel Larner successfully established that the quickest ways to change the mindset of the traditional society was through the use of mass media. He opined that the results could apply to all third world countries as well.

The Indian experience as elsewhere in the world, has been that the media both folk and modern media are inter woven with all aspects of human such as cultural, economic, educational, political, religious, scientific and technological. Because of their speed and efficiency to reach large masses of people, the modern media are employed by our planners and media organisers to share the messages of RD.

In a developing country like India access to communication paves way for modernisation and social change. Being an important factor in the process of development, the mass media not only serve us the media of information but also act as, the agents of social change and pass on new ideas and information.

In India role of print media has limited in RD because of low literacy rate. The effective approach to reach rural masses more information about villages in the print media What the reporters should keep in touch with opinion leaders in the villages who can influence others by the word of month.

The print media plays a vital role in RD, creating awareness among the people towards a better life by informing them about the developmental activities extended to them. AIR is the prime channel for keeping the rural people over the country informed of the govt's plans, policies and programs. It educates rural people on various subjects, provides entertainment, and disseminates knowledge that helps in performance of farm and home operations.

Since the mass media of communication act not only as purveyors of information on but as change agents in many society especially in rural society, the best approach for development would be to use combination of print and broadcast media effectively. Multimedia approach is best to diffuse development themes among rural people.

Journalism and Mass Media

We have been using various means and modes for farm journalism/communication as indicated below

1. *Print Media*: Such as newspapers, magazines, wall-papers, books, posters, folders, booklets.
2. *Audio/Visual Media*: Mainly radio and TV, besides, films, slide shows, bioscope.
3. *Folk Media*: It has been a common observation, that through folk media, it is possible to achieve penetration to the segments where other media have limitations. Puppet shows, street plays skits, stage performances, folk songs and folk dances, nautanki and other traditional means of communication belong to this category. This media is cost effective, entertaining and offers' two-way communication.
4. *Interactive Media*: Science exhibitions, science fairs, seminars, workshops, lectures, scientific tours, conferences, vigyan jathas, etc. The advantage here being man to man and two way communication.
5. *Digital Media*: information technology has given birth to comparatively a new media, known as digital media. It includes Internet, CD-ROM, multimedia, simulations, etc. segments of the society.

That apart, we are popularizing Agriculture through our 18 regional languages, to penetrate into local populace effectively. Selection of target audience has greatest significance. Our farm communication efforts are aimed at various target groups, such as students, farmers, farm women, workers or specialists, etc. Various forms for presentation are being used to making science communication more interesting and enjoyable, such as science news, report, article, feature, story, play, poem, interview, discussion, lecture, documentary, docu-drama, scientoon (science + cartoon), satire, etc.

Today there are popular farm magazines in almost every Indian language, barring a few. Farm programmes appear on radio and TV.

The use of broadcast and digital media has opened new vistas of farm journalism. The revolution in information technology has made possible to get farm information from around the globe within seconds, on our fingertips. More concerted co-ordinate and integrated efforts have started to cater to large and all cross sections of farmers in the society.

Writing Techniques for Print media

The following dimensions should be considered while writing for print media

1. Contents of News
2. Format
3. Style
4. Lead
5. Lay out and Balancing
6. Writing techniques

Contents

The early American newspapers offered information pertaining to births, weddings, deaths, ship arrivals, prime prices for crops, politics, taxes" and related items. Today's newspapers continue to provide such news, but in competing for audience attention against television, cable vision, motion pictures, and magazines, they entered another field entertainment.

The news contents of a newspaper, of course, remains the vital part that persuades the reader to become a regular subscriber. The fringe offering of entertainment becomes larger in the newspapers with bigger circulations and consequently more pages.

The Ingredients of a Newspaper

Aside from advertising, the contents of a newspaper can be broken down into: (1) Entertainment features and (2) News stories. The latter cover a wide range of writing techniques and approaches.

a. Entertainment Features

Mixed with news items throughout a newspaper's are the entertainment features, which are a bonus to readers. These non-news features include comics, crosswords and other puzzles, cartoons, dress patterns, and columns on such specialized interests as chess, pets, astrology, child care and photography.

Large newspapers can afford to have many such features produced by their staff but most dailies buy the material from syndicates that profit by selling to publications throughout the nation.

b. The News Contents

The remaining and basic part of the newspaper of course is news. It can be characterized, categorized, divided and described in a number of ways.

c. Levels of News

The relative importance of news, as decided by reporters and editors, determines whether a particular story is vital enough to be used and where it will appear in the newspaper. Journalists, weighing the value of news situations, subconsciously classify items as "hard" or "soft".

a. Hard News

Stories of primary or urgent importance rate the "hard" news classification. Examples would include articles concerning a key role in the United Nations Congress approving or rejecting controversial legislation, the police department breaking a crime ring, the city council approving a substantial tax increase or decrease, or plans for a facility that would greatly increase the quality of living in the newspaper's community. These situations would provide “must stories” that readers would expect to see.

b. Soft News

News of secondary importance is labeled "soft" news and might concern such items as a school or church social meeting, reports of minor crimes, a city council meeting that produced routine action, or a "human interest" story concerning a pet, hobby personality sketch, or club activity. Such articles would lack great consequence and would be "optional" stories that readers probably would not miss if omitted.

While few people would debate the assignment of the "hard" news category to a tense or controversial international situation, there can be much disagreement over what constitutes important news at other levels. A battle for a baseball pennant can be a "hard news for a baseball fan but of little interest to one who does not follow the sport. A parents' meeting to discuss school problems will be of major importance for those involved in the turmoil but of no interest to those who have no children or reside in another neighborhood.

The referees who ultimately decide what is "hard" or "soft" news are the reporter, who decides what facts are worthy of placing in a story, and the editor, who selects the stories to be published.

c. Editorials and Columns

Two other areas that belong in the news category are editorial and column writing. Editorials are the opinions of the newspaper management regarding situations that have been aired in news stories, and columns are regular contributions by an individual.

The editorial "we," indicating the collective opinion of the newspaper, often is used in editorials. While some columnists prefer to write in the third person, many prefer to use "I" or a poetic "we" to' add a personal feeling to their material.

d. Writing Forms

A single news situation would inspire varied approaches according to writing styles. Here is what might be expected with the announcement that the Soviet premier planned to visit the United States.

The objective news writer would confine the story to the information that the premier planned a visit and would provide available details pertaining to the dates and itinerary.

The interpretative reporter would supplement the basic information of the projected visit with background data drawn from Russian-American experts on what the tour might mean to international relations.

The advocate reporter would write an outspoken news story lauding or condemning, according to personal feeling, the proposed visit.

The editorial writer would produce a contribution weighing the relative values of the trip concluding with an opinion as to whether the visit would or would not accomplish its goals.

The columnist would have free way to handle the situation in a number of ways: a satire of how the premier might tour America as an average traveler recollection of people who traveled to Russia comments on the expectations of the visit, perhaps with background information on tours by foreign officials.

Elements of News

1. Immediacy or timeliness is of great importance. The new thing or latest information makes all the difference. As news ages, it loses its potency. The nearness of an event in time is vital to daily newspapers.
2. Truth and objectivity is another quality of news. The balanced picture of an event or a happening or accuracy in reporting events enhances the value of news.
3. Nearness or propinquity, i.e. closeness in space, is yet another element of news. - The news worthiness of an event is in direct relation to its physical nearness to the public served. An accident or a crime near at home has mush greater news value than a bigger accident or crime abroad.
4. News should be interest or concern or significance to readers. The interest, however, differs. It may be of interest to a section of the community or of local, regional, national or international interest.
5. Names of individuals also make news. The person may be widely known because of his position, wealth or status. Even an unknown person can make news because of some oddity, eccentricity or unusual occupation or some other factor.
6. Timing and place of occurrence of an event are also important elements of news. The hour of an event may decide its news importance. Similarly where the thing happens is also important.
7. Numbers and size make news. An accident in which a large number of people have been killed or injured makes news. If very few people are involved and nobody has been killed the news value becomes less important.

8. Suspense and mystery enhances the value of news. It holds readers' interest in the news.

The other elements which are dominant in news stories are action, movement and mobility; combat, struggle, contrast, rivalry; sex and scandal; beauty and romance, adventure and pathos including exploration, discovery, excavation, invention, novelty humour and stories about children and animals.

In brief, news is associated with something new or something novel or relates to a famous person or of importance to a large number of people or pertains to conflict, mystery, crime or reveals something confidential or forecasts the deals with a funny situation future or romance or sex, suicide, murder, etc.

6.2. FORMAT

1. The paper

The standard paper size of 8 ½ × 11 inches is preferred for stories (Smaller paper sizes are utilized for the cut lines to photographs and for writing headlines; other chapters deal with these specialties). Reporters do not use onionskin or other light weight paper because flimsiness makes it difficult to copy read.

2. Carbon copies

Many newspapers, particularly the larger ones require that the reporter produce at least one carbon of each story. One copy is edited and sent to the type setters, while the other is retained, at least temporarily, while the story is in the mechanical department. Some newspapers require multiple carbons for numerous purposes; the editor, permanent files, reporters related to stories, and for the reporter's reference in handling additional stories on the same subject.

3. The heading

Contrary to widely held public opinion, reporters do not write the headlines to their stories. These headlines are composed by copy readers. Reporters, however do have certain obligations. While "theme" papers for conventional english classes contain introductory material in the upper right hand comer the opposite is true for journalistic copy; it appears, single spaced in the upper left hand comer. These items are as follows:

a. The reporter's last name

If problems or questions develop regarding a story the last name quickly advises the identity of the staff member to be contacted. Since editors, not the reporters, make the decisions regarding which stories shall have by lines, the required name is readily available without someone shouting, "Who wrote this?"

On papers large enough to have more than one reporter with the same last name, the writers obviously would need to identify themselves with initials or first names.

b. The "Slug"

One or two words expressive of a story's contents constitute the "slug" the purpose of which is to help people "follow" the story on its way into print. For example, slugs such as "hotel fire" or "mayor election" indicate the subjects involved when determining whether stories are with editors, copy readers or type setters. Staff members can use the slug words instead of lengthy descriptions. Different machines usually set the smallest type in the body of stories and the larger type in headline so the use of the same slug with the story and the headline serves to bring the two kinds of type together on the proper page in the newspapers.

The reporter's responsibility is to devise the slug being certain that it is reasonably expressive in order to avoid confusion with those that logically could be coined by other writers on the same newspaper. It is important for the slug to be concise and therefore, easily written or used in oral discussions.

Caution

Do not use derogatory, obscene, humorous, or other extraneous or needless words in the slug or elsewhere in the story, assuming that they will be seen ably by staff members. These words, through mechanical or editing errors, could be set in type and result in a libel action if printed.

c. The date

A time element in a story makes it important to know when it was written. Without the date, "tomorrow" or "today" as expressed in the story, could confuse and mislead the reader. The subscriber logically believing what is printed. Hence the date at the top of the story enables editors to make appropriate changes in the article according to when it will appear.

4. Where to begin

After placing the above information on the copy paper, the reporter moves to a point approximately half-way down the page to start the story. This white space serves a number of purposes. It provides room for direction for the style of head line, instructions for the width and size of type, the number of the page in the newspaper where the story will appear, or an editor's questions regarding some fact in the copy.

5. About typing

The type writer is customarily set to provide one inch margins on each side of the paper. While newspapers traditionally have required that copy be double-spaced, many now demand triple spacing. The reasons are obvious. Reporters may wish to add a word or phrase on top of a line. Editors require the space between lines to correct spelling, make style changes. or interject words or phrases they believe will help the story.

6. Brief paragraphing

Beginning journalists usually have backgrounds essential. Involving conventional English compositions, where ideas may develop in paragraphs as long as two pages of double spaced typing. They are therefore, frequently surprised and even disbelieving to learn that paragraphs in newspapers cover approximately four lines of type writing or 40 to 50 words. In fact, some paragraphs are even shorter. Experienced reporters and editors usually regard six lines of type writing is the maximum. There are three reasons for this construction.

a. Readability

People who begin reading a book or magazine often have subconsciously committed themselves to complete their selection, and despite interruptions will finish. Average newspaper reader, in contrast read paper, during spare moments and with interruptions such as drinking coffee, looking from window, while commuting, or conversations. Hence brief paragraphs

- Permit them to return to their reading point without searching through long paragraphs.

- Allow them places to abandon reading if they are not interested in one story and wish to skip to another.

b. Ease in typesetting

If the typesetter makes the mistakes of dropping words or phrases, he needs only to return and reset a few lines of type when the paragraphs are short. If the paragraphs are lengthy and the errors occurred early in them, it probably would be necessary to reset a substantial part of the story to include the missing words. Despite care, typesetting mistakes are made, and newspapers seek to keep costs down by correcting them as easily as possible.

c. Making up the newspaper

Fitting stories on top of advertisements, around photographs and other articles is an art. Printers and the news department's make-up editors work together on this problem. Dropping a few lines from a story often will make it fit the desired space on a page, and it is easier to delete one paragraph with a few lines than to break into a large one. Cutting lines from a large paragraph usually requires extensive resetting of type, thus boosting time and costs.

7. Splitting words, sentences and paragraphs

Journalists avoid hyphenating words at the end of lines because typesetters, working speedily are slowed by hyphens, which they may set almost automatically in the center of a line. To avoid hyphenation, reporters either stop short of a full line if word may not fit in it entirety or press the margin release to complete the word.

8. Stories with more pages

Journalists use several techniques to achieve clarity with stories consisting of more than one page.

a. Ending pages

On reaching the end of a page, the writer types "more" at the bottom to advise editors or typesetters that there is more to the story. Unlike fiction, news stories may logically end at various points, without indicating that there is "more" the reporter could cause others to stop after the first page and not handle the balance of the article.

b. New pages

Unlike the first page, the second starts approximately an inch from the top. Starting in the upper left with single spacing, the reporter types his last name and the same slug used on the first page. He then types "Add 1” meaning that this page is the first addition to the first page. Some reporters prefer simply to use “Page 2”. The story is double or triple spaced in the same fashion as the first page.

c. End of the story

When the reporter reaches the conclusion of an article he shifts to a new line and in its center types "30" the symbol made so famous on TV shows and fiction relating to journalism. It traditionally ends stories so that editors and typesetters will not be expecting more pages. Researchers debate the origin of the symbol, tracing it variously from telegrapher's terms in the 1800s to phrases used in printing. Regardless of how it originated "30" is used almost universally in journalism as an abbreviation for "The End".

9. Unusual names or spellings

News stories often involve situations providing unusual or unfamiliar spellings. Instead of the traditional "Smith" the name may be "Smithe" or it may be "Briwn" rather than 'Brown'. Without the reporter present, conscientious copyreaders may debate whether the reporter hit the wrong key or was accurate in spelling an unusual name. The reporter should verify unfamiliar spellings by typing "ok" over the word placing it after on the same line. An alternate to "ok" is "cq", derived from an' early. telegraphic term meaning that the transmission was correct.

10. Unusual or incorrect material

To produce an effect, a reporter may use old English phrases that contrast present day spelling and usage, or even material that is grammatically incorrect, or writers also may quote from literate ramblings. These situations call for alerting by an editor or typesetter to make corrections which would ruin the effect. To avoid this, the reporter writes "follow copy" on the story, thus indicating that the unusual writing is purposeful and not intended to be corrected.

The Latin term "sic", meaning that what might sound surprising or incorrect is true, is seldom used in newspapers because of the belief' that the average would not understand the term.

6.3. Style

a. Definition

"Style" is defined in the dictionary as the "manner of dealing with spelling, punctuation and word division of a particular press.

b. Style book

"Style" for newspapers, magazines, and books extends past spelling to capitalization, abbreviation, punctuation, the form of titles, and whether numbers will be figures or spelled. These matters are established by editorial executives and distributed to the staff as a "Style book".

c. Points to be considered in style of writing

Every news agency or newspaper follows its own 'style'. But a few general guidelines in regard to the style of' writing are given below:

1. Use short words, short sentences and short paragraphs.
2. Produce a clean copy.
3. Avoid needless adjectives so that the risk of exaggeration or under estimation is eliminated.
4. Distinguish between factual and propagandistic material in what is available from your sources whether official or non-official.
5. Guard against word wasting abstractions.
6. Where there is no risk of ambiguity or awkwardness, drop the point from groups of initials such as CPI, SP, IAF, PIB and AIR. But where confusion may occur, use points as in

U.N. (United Nations), U.S (United States), W.H.O. (World Health Organisation).

7. Do not abbreviate names, say Mohammed (not Mohd.), Chaudhary (not Ch.).
8. The standard abbreviations are:
   - kg, km, kw, kv (for both singular and plural);
   - Mr, Mrs, Dr. (Preceding a name) Jr. Sr (following a name);
   - Academic degree: B.A., M.A., M.S., Ph.D., M.D.;
   - M.P. (Member of Parliament)
   - GMT (Greenwich Mean Time), IST (Indian Standard Time), Hrs" (hours)
   - I.P. C. (Indian Penal Code)
9. Do not abbreviate the days of the week and the months of the year in the body of the copy.
10. Similarly, say Rupees, not Rs. except in tables.
11. Do not use i.e. (id est) for that is, e.g. (exempli gratia) for example; etc. (et cetera) for and so forth.
12. Use the following abbreviations for service ranks:

Lieutenant - Lt	Lieutenant General Lt Gen
Lieutenant Commander - Lt. Cdr	General - Gen.
Commander – Cdr	Flight Sergeant - Fl t Sgt
Captain - Capt	Flight Lieutenant - Fit Lt
Captain - Capt	Squadron Leader - Sqn. Ldr
Major - Maj	Wing Commander Wg. Cdr
Lieutenant - Colonel - Lt. - Col	Group-Captain - Group-Cpt
Brigadier - Brig	Commodore – Cmdr
Major General - Maj Gen	Warrant Officer -WO.

13. In the first instance the rank should be given in full.
14. Except May, June and July, the months are to be abbreviated in dateline: JAN, FEB, MAR, APR, MAY, JUNE, JULY, AUG, SEP, OCT, NOV, DEC.
15. The months are, however, not abbreviated in the body of an item nor the days of the week.
16. Say October 30 and not 30th October as a matter of style.

17. Calendar years to be mentioned in full thus : 1966 (not' 66) and: financial year 1976-77.
18. The days of the week and the months are not to be abbreviated in the body of a story.
19. Except in quotes, the time should be rendered on the basis of 24 hour cycle. Say 11:15 hrs and 21:30 hrs.
20. GMT time should be converted into IST and both should be given occurred at 11:00 hrs IST (05:30 hrs GMT).
21. All editing should be done in capital letters only. Heavily edited copy be retyped.
22. Type 'no' and 'now' twice. Also type twice uncommon names of persons and places, Example: "He said now now." This is to ensure "now" does not appear as 'no' in transmission, which happens quite often.
23. Numbers up to eleven should be spelt out and those above given in figures.
24. For money use lakhs and crores, not millions.
25. However, for expressing the number of men and other things use millions. For example: A grant of rupees five lakhs ... but 5,000,000 people (in figure) six million shoes 96.50 million tones of food grains.
26. Except when confusion is likely to result the following are not put in quotes: names of newspapers or magazines; trade names makes or models of cars and aircraft; sports events or trophies; fictional or stage characters ship or pets, names of royal, official or historical houses and castles.
27. The following are normally put in quotes: nicknames or pen names; titles of books, poems, songs, plays and films; private houses or properties in general.
28. Avoid abbreviations like e.g., n.b., viz, etc. and use their equivalents: for example, note carefully; namely and so forth.
29. Do not use in the copy trade names for the products of the same type manufactured by others. Say a detergent powder (surf); transparent. Wrapping (cellophane); vacuum flask (thermos); synthetic fabric (terrine); petroleum jelly (Vaseline); aspirin (aspro) and edible oil (vanaspati).

11. Art of Speech

11.1. What is oratory?

Buehler and Johannesen defined oratory as "a memorized, original, persuasive speech, dealing with worth-while subject matter of timely interests, demonstrating qualities of logic, organization, language and delivery, and producing an effect of eloquence which is far above the ordinary."

Oratory uses the principles of the art of **rhetoric**. Rhetoric is one of the oldest disciplines that are still actively studied around the world.

11.2. The art of Rhetoric

Aristotle was one of the first to write about rhetoric in his book *Rhetoric*. In this book he defines rhetoric as "the art of finding, in every given case, all the available means of persuasion." Basically, rhetoric is the art of making people believe what you want them to believe by using any means of persuasion. Rhetoric provides many principles by which to write, present yourself, to be trusted, and to be logical. You will learn about these principles throughout the site. The first orators were called sophists. They spoke on political issues and also taught others how to speak. Oratory is a critical support to democracy. As the art progressed, it was integrated into the church. Today it is found primarily in the government but also with our religious figures, actors, and other public speakers.

11.3. Principles of Rhetoric or oratory

Rhetoricians defined five canons or principles of rhetoric and they are invention, arrangement, style, memory, and delivery. Each one of these canons has its special purpose and not all of them are used in every oratory.

a. Invention

Invention is the canon that is used to develop material. Sharon Crowley defined invention as "Invention becomes the study of all the possible means by which arguments or proofs can be discovered and developed." Lawrence J. Prelli defined as "A rhetor must be able to discern what is at issue within a rhetorical situation and how the content is pertinent to the issue and to discover content suitable for rhetorical intervention”

b. Arrangement

Arrangement is the organization of thought, principle, and evidence in order to make a speech, persuasive. Socrates wrote that "Every discourse, like a living creature, should be so put together, that it has its own body and lacks neither head nor feet, middle nor extremities, all composed in such a way that they suit both each other and the whole."

Oratory has seven parts and they are

1. The entrance, or the introduction of the subject and the securing of the subject at hand
2. The narration, situations necessary to understand the topic
3. The proposition, the speaker's central idea or thesis
4. The division, or a brief list of the points the speaker will demonstrate
5. The confirmation, or the body of proof for the points
6. The confutation, or the rebuttal
7. The conclusion

These seven parts are used loosely in modern oratory.

c. Style

Style is the third of the five canons. Style is the soul of a speech - used to grab hold of listeners. If there were not style, a speech would be just words, which had no higher value.

Style also means the proper use of language, is critical to the listeners and their understanding of the speech. Aristotle stresses this concept and writes to the point that if your listener does not understand your language, they cannot understand your speech.

d. Memory

Memory is the fourth of the five canons. Memory refers to the listeners' ability to remember what the speaker has said. If the speaker properly present the other four cannons, this canon will be achieved.

e. Delivery

Delivery is the way in which the speaker present the speech. It embodies everything from the way the speaker stand, to how the speaker speak and pronounce each and every word. Delivery is what links the other four canons together in the speaker 's presentation.

11.4. Parts of Speech

The speech has three parts viz., introduction, body and conclusion.

a. Introduction

- The introduction should be direct and quick.
- Arouse curiosity by giving some factual statements.
- Begin with an illustration.
- Shocking facts have great power to make the audience attentive.
- Never begin with an apology.

b. Body

- Body is the main speech.
- In the main speech as per planning talk over the identified points.
- At the end of the meeting these points should be remembered by the audience.
- Eradicate all ill feelings.
- Don't challenge.
- Give local and suitable examples.
- Support the points with adequate evidences.
- Use simple and local language.
- Be time conscious.
- Acknowledge the organizers and audience.
- If possible, use appropriate visual aids. Visual aids help the audience to remember longer.
- Quote relevant proverbs and quotable quotes.
- Restate your ideas in different words.

- Wherever possible repeat the points.
- Try to avoid contradictory ideas.
- Ideas should be presented with enthusiasm and conviction.
- Quote authority to support your statements or points.

c. Conclusion

- In any speech the introduction and conclusion are very important.
- The last part of the talk is conclusion.
- Closing is very important because the final words are remembered longer.
- The closing should be smooth.
- It should be done by summarizing the main points, with an appeal for action.
- If possible close with humour and also with some practical questions based on the nature of the subject.

11.5. Elements for effective speech

1. Be Prepared

- The topic should be need based.
- Don't try to cover a large ground.
- Converse on the topic.
- Collect your experiences for illustration.
- Empathize in the place of audience, what is needed to them and what would interest them.
- After collecting the thoughts select the useful ones and discard those which would not benefit the audience.
- Read extensively and have additional information at reserve for answering to the audience doubts.
- Plan the speech, write it and rewrite after polishing.
- Remember the main points, if necessary write them on a slip of paper
- Being prepared is by far the most important element. How many times do you practice your speech? As a general rule, you should spend about 30 hours of preparation and rehearsal time for every hour you will be speaking.

- Use a tape recorder or video tape yourself. This will help you to get an accurate picture of how you speak.

2. Give of Yourself

- Use personal examples and stories in your speech whenever possible.
- Make sure your stories help to emphasize or support your point.
- The stories must match your message.
- Use examples from your personal and professional life to make your point.
- In either case be willing to give of yourself by sharing some of yourself with the audience.

3. Stay Relaxed

- To stay relaxed you should be prepared.
- Focus on your message and not the audience.
- Use gestures, including walking patterns.
- Practice the opening of your speech and plan exactly how you will say it.
- The audience will judge you in the first 30 seconds they see you.

4. Use Natural Humor

- Don't try to be a stand up comedian. Use natural humor by poking fun at yourself and something you said or did.
- Be sure NOT to make fun of anyone in the audience.
- People will laugh with you when you poke fun at yourself but don't over do it.

5. Plan Your Body and Hand Positions

- During the practice of your speech look for occasions where you can use a gesture.
- Establish three positions where you will stand and practice not only how to move to them but where in your speech do you move.
- Pick three positions, one on center stage, one to your right, and one to your left. Do not hide behind the lectern.

- When you do move maintain eye contact with the audience.

6. Pay attention to all details

- Make sure you have the right location (school, hotel, room and time).
- Make sure you know how to get to where you are speaking.
- Ask how large an audience you will be speaking to.
- Make sure you bring all your visual aids and plenty of handouts.
- Arrive early so you can check out where you will be speaking and make any last minute adjustments.
- Go to the facility early to make sure you're comfortable in the surroundings. Check the microphone, lighting, audio/ visual equipment.
- It is very important that you pay attention to even the smallest details.
- You can never over plan.
- Remember, "He who fails to plan is planning for failure"
- Write your own introduction and make it pertinent, emphasizing your credentials.
- Make sure you absolutely know who is going to be in the audience, why they are there and why they invited you to speak.
- The first 30 seconds have the most impact. Launch right in with a startling statement, quote or story. End with a bang, not a whimper.
- Do not start a speech with humor unless you are absolutely brilliant at it. If you tell a bad joke, you're going to lose any credibility you have. Moreover, if your only humorous material is at the beginning, the audience is going to be disappointed when you become serious.
- If you're giving a 1/2 hour speech, don't expect to tell the audience everything you know. Pick two or three important points.
- If your presentation involves statistics and analytical data, put them in a handout that the audience can refer to. Don't bore them by reciting a plethora of numbers.
- Don't read your speech.

- Look the audience in the eye.
- Write down key points or statements.
- Deliver the rest extemporaneously.

11.6. Other points to be considered for effective speech

a. Dress

- The dress should be neat and appropriate to the occasion.
- The dress should be acceptable to the audience.

b. Stage mannerism

- Sit calmly in the stage.
- Be relaxed and cool.
- Don't talk among your friends in the stage. That will distract the attention of the audience from the speaker.

c. Mike mannerism

- Don't hold the mike with your hand.
- Don't talk too loudly to the mike.
- Don't yawn in front of the mike.
- Tilt your head gently while you want to see the audience seated in the side rows.

d. Gestures

- The movement of head, a wave of the arm, raising the eyebrows, turning the head etc., are all gestures.
- The gestures should be spontaneous and give life to the talk.
- Don't over do gestures
- Gestures which go beyond certain limit are called gesticulation.

e. Eye contact

- It is looking squarely at the people in the audience.
- The eyes sweep round from person to person, section to section forward and back, and side to side.
- Look at the audience and make them feel that you are interested in them.

- It establishes rapport with them.
- It devotes confidence in the speaker.
- It prevents any deviation on the part of audience.
- It keeps the audience attentive all the way.

f. Voice

- A good voice with proper modulation is an asset for good public speaker.
- By regular practice this can be attained.

g. Pause

- The interval between words, phrases, sentences, paragraphs, ideas, section etc., is called a pause.
- It is not possible to speak without a pause.

STYLE AND STYLE BOOK

STYLE

"Style" are defined by the dictionary as the "manner of dealing with spelling

Punctuation and description particular property"

Style Book

"Style" for newspaper, magazines and books extends past spelling to capitalization, abbreviation, punctuation, the form of titles, and whether numbers will be figures or spelt. These matters are established by editorial executives and distributed to the staff as a "style book".

Style

Style is the way things are done around the newsroom. Style is the way you are supposed to do things. It is arbitrary, unreasonable and confusing. But it is important. All newspapers follow a style book of one kind or another. Many use the Associated Press

Stylebook and Label Manual

Style is basically a preference for one way of doing thing over another way when there are two or more acceptable ways of doing them. It is the sense of providing rules for spelling, punctuation, capitalization, abbreviation and the use of members.

Style talks you how you ought to use the language if you are to conform to what have came to be regarded as standard practices among experienced and well educated writers and editors.

Style and policy

Style books usually include some rules or guidelines to the newspaper policy on such matters as courtesy titles, racial descriptions, sexual stereotyping, the use of epithets, vulgarities and use of legal, technical and scientific terminology. The AP style book goes into considerable detail on courtesy titles. Mr. MS. Mrs. & Ms.

The authority behind style

The AP stylebook bases its rules on spelling and usage on Webster's new world dictionary of the American Language and well - edited and upto date dictionary.

Military titles and designation are based on the official usage of the armed forces, and rules on religion are based on the denominations in the United States.

Mechanical rules

The news style also included the mechanical aspects of punctuation, capitalization, abbreviations and the use of numbers. The style book provides helpful guides to consistency in certain areas.

Capitalization

All are agree that the initial letters of personal names should be capitalized and the first letter of the first word of a sentence is capitalized and some other matters are Capitalize formal titles "when they are used immediately before one or more names Pope Paul, President Washington. Capitalize common news such as party, river, street and west when they are all integral part of the full name for a person, place or thing. Democratic Party, Mississippi River, Fleet Street, West Virginia.

Abbreviation

1. AP style basic allows U.S. as an objective but not noun Sets abbreviations of state names, legislative titles,

military titles and religious titles. Numbers are a problem, but style books have one generally accepted rule: write out numbers one through nine and use Arabic Figures for 10 and large numbers. Use figures for amounts under 10 in dimensions, formulas and speeds. The farm measures 5 miles by 4 miles. The car slowed down to 7 mph spell out causal expressions. Thanks a million. News style is gradually eliminating unnecessary punctuation as a means of keeping sentences. It does away with apostrophe in plurals of numbers and multiple letters. The airline has no. 727s in the 1950s.

Built in style books

Many newspapers now have the AP style book or their own style book online, that is, in the newspapers computer system as part of the word processing programme used in writing news stories. This makes it easy to refer quickly the style book while working at the key board.

Personal style

It presents some guidelines for clear and accurate writing. But it is not inhibit you from developing your own style of writing. Personal style is choice your choice of words idioms, sentence structure and grammatical alternatives.

Points to be considered in style of writing

- ❖ Use short words, short sentences, short paragraphs and produce a clean copy.
- ❖ Avoid needless adjectives so that the risk of exaggeration or under estimation is eliminated.
- ❖ Guard against word wasting abstractions.
- ❖ Where there is no risk of ambiguity or awkwardness, drop this point from groups of initials such as CPI, SP, IARF, PTB ad AIR. But, where confusion may occur, use points as in U.N. (United Nations), U.S. (United States), W.H.O (World Health Organisation)
- ❖ Do not abbreviate names, say Mohammed (Not Mohd Chaudhary
- ❖ The standard abbreviations are: kg, km, kw, kv (for both singular and plural); Mr, Mrs, Dr. (preceeding a name) Jr. Sr. (following a name);

- Academic degree : B.A., M.A., M.S., Ph.D., M.D., M.P. (Member of Parliament)
- GMT (Greenwhich Mean Time), IST (Indian Standard Time), Hrs. (hours).
- I.P.C. (Indian Penal Code), Cr P.C (Criminal Procedure Code).
- Similarly, say Rupees, not Rs. except in tables.
- Except May, June and July, three months are to be abbreviated in datelines; JAN, FEB, MAR, APR, MAY, JUNE, JULY, AUG, SEP, OCT, NOV, DEC.
- The months are, however, not abbreviated in the body of an item nor the days of the week.
- Say October 30 and not 30th October as a matter of style.
- Calendar year is to be mentioned in full: 1966 (not '66) and financial year thus: 1976-77.
- The days of the week and the months are not to abbreviated in the body of a story.

Except in quotes, the time should be rendered on the basis of 24 hour cycle. Say 11:15 hrs and 21:30 hrs.

All editing should be done in capital letter only. Heavily edited copy be retyped. Numbers up to eleven should be spelt out and those above given in figures. For many use lakhs and crores, not millions. However, for expressing the number of men and other things use millions, for example: a grant of rupees five lakhs but 5,00,000 people (in figures) ...six million shoes .... 96.50 million tonnes of food grains"

Chapter – 3

Functions of Journalism

IDEALS AND FUNCTIONS OF JOURNALISM

India is the largest democracy in the world with one billion crore (100 crore) people. In the context of large-scale economic development in the industrial and agricultural sectors, the need for communication with the people through different media has assumed great importance. Journalism in India has made significant strides during the 50 years since Independence. It has played an important role in disseminating information and ideas and in providing instruction and entertainment to put people in a purposeful and effective manner.

- ❖ Journalism exists in a number of media; newspapers, television, radio, magazines and, most recently, the World Wide Web through the Internet.
- ❖ The subject matter of journalism can be anything and everything, and journalists report and write on a wide variety of subjects
- ❖ Journalists are usually expected and required to go out to the scene of a story to gather information for their reports, and often may compose their reports in the field.

Role of journalism in society

- **Lippmann** understood that journalism's role at the time was to act as a mediator or translator between the public and policymaking elites.
- **Dewey** believe that journalists should take in the information, then weigh the consequences of the policies being enacted by the elites on the public. Over time, his

idea has been implemented in various degrees, and is more commonly known as "community journalism."

The Elements of Journalism

- Journalism's first obligation is to the truth.
- Its first loyalty to the citizens.
- Its essence, is discipline of verification.
- Its practitioners must maintain an independence from those they cover.
- It must serve as an independent monitor of power.
- It must provide a forum for public criticism and compromise.
- It must strive to make the significant interesting and relevant.
- It must keep the news comprehensive and proportional.
- Its practitioners must be allowed to exercise their personal conscience

Three sources of "duties" in journalism

- General ethical duties: being truthful, keeping promises, avoiding harm and serving the public good.
- Journalism's social role: Like all professionals, journalists use their skills to fulfill a social role(s) and to meet public expectations.
- Impact and influence: Even if journalism lacked a special social role, journalists would incur ethical responsibilities due to their impact on the individuals and groups they report on, and on the society they serve.

Professional and ethical standards

Since the development of professional journalism at the beginning of the 20th Century, journalists have been expected to follow a stringent code of journalistic conduct that requires them to, among other things:

- Use original sources of information, including interviews with people directly involved in a story, original documents and other direct sources of information, whenever possible, and cite the sources of this information in reports.

- Fully attribute information gathered from other published sources, should original sources not be available (to not do so is considered plagiarism; some newspapers also note when an article uses information from previous reports).
- Use multiple original sources of information, especially if the subject of the report is controversial.
- Check every fact reported.
- Find and report every side of a story possible.
- Report without bias, illustrating many aspects of a conflict rather than siding with one.
- Approach researching and reporting a story with a balance between objectivity and skepticism.
- Use careful judgment when organizing and reporting information.
- Be careful about granting confidentiality to sources (news organizations usually have specific rules that journalists must follow concerning grants of confidentiality).
- Decline gifts or favors from any subject of a report, and avoid even the appearance of being influenced.
- Abstain from reporting or otherwise participating in the research and writing about a subject in which the journalist has a personal stake or bias that cannot be set aside.

Functions of journalism Utilitarian

1. Conveys national policies to the public
2. Keep the government at local state and central levels informed of public needs
3. Brings to the notice of the government on public reaction to Government Policies and decisions.
4. Keeps the public and the government informed of events and happenings at home and abroad.

Entertainment

It provides entertainment to its readers by publishing short stories, poems, sports, cinema features etc."

Responsibilities of journalist

Social responsibilities

- Journalism is the mirror of the society.
- Spokes person for the mankind
- Facts and opinions must be sieve by objectivity

Legal responsibilities

- Safeguard from legal clutches
- well aware of press laws
- within the limits of the law and never infringe them

Professional responsibilities

- Commitment to his profession
- Know the limits within which to operate
- Let news not be distorted, suppressed or exaggerated to suit anybody's personal interests
- Interest of the readers should be taken account. Avoiding of errors in the language is as much essential as avoiding of errors in the content.

Journalistic idealism

- Ethics is a branch of philosophy that helps journalists determine what is right in journalism. It is like self-legislation and self enforcement.
- Journalists as one is not simply writing for the education of others. He puts himself and his ideas into his journalism.
- The journalist who has concern for ethics does care about or right action.

Merril - A lack of commitment and consistency among journalists. Before any journalist choose any particular ethics, he has certain social considerations in his mind. Generally, the biases are both conscious and unconscious.

CANONS OF JOURNALISM

The primary function of newspapers is to communicate to the human race what its members do, feel and think. Journalism, therefore, demands of its practitioners the widest range of intelligence of knowledge and of experience, as well as natural and trained powers of observation and reasoning. To its opportunities as a chronicle are indissoluble linked its obligations as teacher and interpreter.

The first amendment, protecting freedom of expression from a bridgement by any law, guarantees to the people through their press a constitutional right, and thereby places on newspaper people a particular responsibility, thus Journalism demands of its practitioners not only industry and knowledge but also the pursuit of a standard of integrity proportionate to the journalist's singular obligation. To this end the American Society of Newspaper editors sets forth this Statement of Principles as a standard encouraging the highest ethical and professional performance.

To the end of finding some means of codifying sound practice and just aspirations of American journalism, these canons are set forth:

Canons means: general rules and principles of journalism or an authentic set of works. Even when there is no legal authority to enforce canons, 1st as a profession, is supposed to put deliberate ponderings to personal instincts in born tenting. If canons are not followed it can yield public disapproval and there by professional condemnation.

1. Responsibility

The right of a newspaper to attract and hold readers is restricted by nothing but considerations of public welfare. The use of newspaper makes of the share of public attention it gains serves to determine its sense of responsibility, which it shares with every member of its staff. A journalist who uses his power for any selfish or otherwise unworthy purpose is faithless to a high trust.

The primary purpose of gathering and distributing news and opinion is to serve the general welfare by informing the people and enabling them to make judgments on the issues of the time. Newspaper men and women who abuse the power of their professional role for selfish motives or unworthy purposes are faithless to the public trust. The American Press was made free not just to inform or just to serve as a forum for debate but also to bring an independent scrutiny to bear on the forces of power in the society including the conduct of official power at all levels of government.

2. Freedom of the press

Freedom of the press is to be guarded as a vital right of man kind. It is the unquestionable right by laws including the wisdom of any restrictive statute. To its privileges under the freedom of American institutions are inseparably joined its responsibilities for an intelligent fidelity to the constitution of the United states.

Freedom of the press belongs to the people. It must be defended against encroachment or assault from any quarter, public or private, Journalists must be constantly alert to see that the public's business is conducted in public. They must be vigilant against all who would exploit the press for selfish purposes.

3. Independence

Freedom from all obligations except that of fidelity to the public interest is vital. Promotion of any private interest contrary to the general welfare, for what ever reason, is not compatible with honest journalism. So called news communications from private sources should not be published without public apply to so called special articles unmistakably devoted to advocacy or characterized by a signature authorizing the writer's own conclusions and interpretations.

To be impartial does not require the press to be unquestioning or to refrain from editorial expression. Sound practice, however, demands a clear distinction for the reader between news reports and opinion, Articles that contain opinion or personal interpretation should be clearly identified.

4. Fair Play

A newspaper should not publish unofficial charges affecting reputation of personal characters, without opportunity given to the accused to be heard' right practice demands the giving of such opportunity in all cases of serious accusation outside judicial proceedings.

- ❖ A newspaper should no invade rights of private feelings without sure warren of public right as distinguished from public curiosity.

- ❖ It is the privilege, as it is the duty, of a news paper to make prompt and complete correction of its own serious mistakes of fact or opinion whatever their origin.

5. Decency

A newspaper cannot escape conviction of insincerity, if while professing high moral purpose, it supplies incentives to base conduct are to be found in details of crime and vice, publication of which is not demonstrably for the general good. Lacking authority to enforce its canons, the journalism here represented can but express the hope that deliberate pandering to vicious instincts will encounter effective public disapproval or yield to the influence of a preponderant professional condemnation.

These principles are intended to preserve, protect and strengthen the bond of trust and respect between American journalists and the American people, a bond that is essential to sustain the grant of freedom entrusted to both by the nation's founders.

Adopted in 1922 as the "Canons of Journalism' Revised and, renamed "Statements of Principles" in 1975.

ETHICS OF JOURNALISM

Ethics is a branch of philosophy that helps journalists to determine what is right in journalism?

- Self determined/ voluntary

- Like self legislation and self enforcement.

Ethics of journalism Ahuja B.N. 2004 by the words of plato in:

(i) Wisdom (ii) Courage (iii) Temperance (iv) Justice (v) Truth.

Gives direction to the moral life

Intellectual base for any system of ethics

Wisdom —> acquired knowledge native abilities

Source —> Experience, from contemplation reading, conversing and study.

- ❖ Helps to resist the many temptations
- ❖ Keeps one constantly pursuing his goal. eg. investigative journalist.
- ❖ demands reasonable moderation
- ❖ gives harmony and proportion to moral life

- ❖ helps us to avoid fanaticism in pursuit of any goal
- ❖ refers to man's social relations
- ❖ considering a mans "deservingness"
- ❖ Authenticity of the information contained

The ethics of collective altruism

It has been expressed generally in terms of the **utilitarian principle** that good conduct is that which results in the greatest good to the greatest number.

Practical problems with this theory;

1. The problem determining, what is really good for most people.
2. The problem possessed by equality "good" with majority opinion or action

Codes of Ethics in **India**

1968: The All-India Newspaper Editor's Conference (AINEC) - formulated code of ethics

1976: Parliamentary code

1991: International code

AINEC CODE OF ETHICS (1968)

(1) A free press

- Flourish only in free society
- Communalism in a threat to fabric of our free society
- Hindrance to national solidarity

(2) Consummation of fundamental objectives

- Democracy, secularism, national unity and integrity enshrined in our constitution.

(3) Guidelines in reporting on communal incidents in the country

a. Freedom scurrilous attack against leaders / communities

b. Generalised allegations casting doubts and aspersions on the patriotism and loyalty of any community should be eschewed.

c. Communal hatred and distrust must be avoided

d. Truth should not be suppressed

e. News of incidents involving loss of life, Lawlessness arson etc. should be described reported and headlined with restraint in strictly objective terms and not in prominent.

f. Peace and Harmony related news items should be given prominence and importance

g. Caution should be exercised in selection and publication of pictures/ cartoons / poems etc.

h. Names of the communities should not be mentioned (majority / minority communities be ordinarily)

i. Sources should always be indicated

j. Verification of facts / figure, before publication is very essential.

Press council guidelines

a. Provocative and sensational headlines should be avoided

b. Headings must reflect and justify the matter printed inside

c. Figures of casualities given in headlines should be preferably be on the lower side.

d. Quotation marks should be used in the case of statement containing allegations.

e. Comments and value judgement in presentation of news should be avoided

f. Presentation of news should not be motivated or guided by partisan feelings

g. Language employed in writing the news should be temperate

h. Corrections should be promptly published with due prominence and regrets.

Press council code on communal writings

It deals with code of ethics on communal writing. It lays down what the press should avoid.

1. Distortion or exaggeration of facts or incidents in relation to communal matters
2. Employment of intemperate or unrestrainted language in the presentation of news.
3. Encouraging or condoning violence
4. Exaggerate real grievances.
5. Scurrilous and untrue attack on communities
6. Falsely giving a communal colour to incidents
7. Emphasising maters that are apt to produce communal hatred.
8. Publishing alarming news which is substance untrue
9. Exaggerating actual happenings to achieve sensationalism
10. Making disrespectful, derogatory, consulting remarks.

The parliament code

Adopted in 1976 a code of ethics for journalists and newspapers in pursuance of its social and moral responsibility.

- In the discharge of their duties, journalists shall attach full value to fundamental human and social rights
- Highlight activities of the state and public, promote national unity, solidarity integrity and economic and social progress
- Avoid reports and comments which tend to promote tension.
- Ensure that information dissemination is factual
- No sensational report shall be published
- Professional secrecy shall be preserved
- Inquiries shall not allow personal interest
- There is nothing so unworthy as the acceptance of a bribe for the exercise by a journalist of his power to give or deny publicity to news.
- Not indulge in personal controversies in which no public interest is involved
- Shall not give currency to public rumours or gossip
- Shall refrain from publishing matter which is obscene

- Shall promote and project the national objectives of democracy, socialism and secularism
- Shall refrain giving tendentious treatment to news of disturbances involving caste, community, class, religion
- Shall not publish information and comment detrimental to the interest of the sovereignty.

The international code of ethics

It was drafted and adopted in 1991 at **Stockholm symposium.** Various points in the code are as follows.

- The fundamental objective of a journalist is a fair accurate, and unbiased story.
- Unnamed sources shall not be used
- Journalists have a moral obligation to protect confidential sources of information
- Corrections or clarifications should be published for errors of fact promptly
- Journalists should identity themselves, except in some exceptional cases
- The use of any obscene or tasteless language should be limited to quoted material
- In principle, journalists should avoid paving for information unless public interest is involved,
- Plagiarism i.e. using someone's work without attribution, is a serious ethical breach and it should be avoided
- In general, race of the person in the news should not be reported,
- Except in rare and justifiable circumstances, Journalists should not tape anyone without that person's knowledge,
- Privacy of an individual should always be safeguarded except in some exceptional cases
- Sex discrimination should not be done
- Journalists should avoid offending individual and yet not compromise their reporting of the news
- Children should distinguish clearly between comments, conjectures and facts

- Living victims of sex crimes should not be identified in the news story
- Journalists should judge no one unheard
- Journalists should maintain highest professional and ethical standards

RESPONSIBILITIES OF JOURNALISTS

The press in no doubt an industry, a business, but with a difference. What is published in newspapers and magazines, conditions the minds of readers: men and women, children, adolescents and adults. Journalists, therefore, have some social obligations. They must act with a profound series of responsibility towards the society.

- ❖ The responsibilities of a journalist are many and varied. His primary duty is to serve his readers through fair, positive and timely information in a palatable, digestible and entertaining format.
- ❖ He must give special attention to the deprived and the weaker sections of society and try to involve them in the country's process of development.
- ❖ He must act as a harbinger of social change.

Social responsibilities

The social responsibility of Journalists is of paramount importance. It is now well recognized that "journalism is the mirror of society". The journalist acts as a spokesman of mankind. He must therefore, provide a truthful, comprehensive, and intelligent account of the events in a context that gives meaning. Facts and opinions must be clearly differentiated, by that effective sieve called "objectivity".

The press must serves as a forum for the exchange of comment and criticism. The Journalist should be an effective medium for communication between leaders and different organizations of the society.

Legal responsibilities

Publication of baseless, graceless and manipulated material should be avoided at ill costs in order to safeguard yourself and your organization from the legal clutches. 'Libel" or "defamation" is always a "live wire" on which a sensible journalist dreads to lead. He must be well aware of the various laws relating to the press such as Official Secrets Act, Copyright Act, he must always act within the limits of the law and never infringe them.

Professional responsibilities

- The basic responsibility of journalist is to disseminate information in all circumstances, sometimes normal and sometimes otherwise. The information provided helps the readers to make up their minds on vital issues. A journalist is expected to have his commitment to his profession. He must strictly adhere to the accepted professional norms.
- He has to know the limits within which to operate. Whether working as a reporter, sub editor, editorial writer or whatever else, he must keep his biases at bay and should endeavour to be fair and balanced in the projection of news and views. Let news not be distorted, suppressed or exaggerated to suit anybody's personal interests - not his own, not his proprietor's not of any business group, political party or government. Similarly, in the projection of views and comments, only the interests of readers at large, of the community and the country, should be taken into account.
- In modern times, news is a business a competitive business. At times, this profit making exercise results in hasty, biased, ambiguous and sensational responding. A journalist just avoid such misdemeanor and keep his integrity intact. He must never jump to premature and incorrect conclusions.
- The "opinion polls, much in vogue these days, are certainly an exercise in premature judgement and they do interfere in the electoral process".
- One of the most important responsibilities of a professional journalist is to ensure that in hurry and haste, accuracy in the language is not sacrificed.
- Avoiding of errors in the language is as much essential as avoiding of errors in the content.

Editor's responsibilities

1. Under the law, the editor has to assume responsibility for all matter published in the newspaper: news, views and

advertisements. An eight point Editor's Charter adopted by All India Newspaper Editors' Conference (A1NEC) in 1953, is still valid. It is as follow:

2. The Editor shall enjoy complete freedom in respect of the implementation of the editorial expenses and the staffing and conduct of the paper.
3. The board of management of a newspaper shall prepare and annual editorial expenses in consultation with the editor, providing the normal expenditure, development programme, contingencies and discretionary grant.
4. The Editor's decision shall be final in all matters concerning the editorial staff and the contents of the paper.
5. The Editor shall have direct access to the Board or to the proprietor for discussion of matters relating to his department.
6. The code of obligations for the Editor is to carry out the policy of the paper, to maintain high standards, to resist all pressures to generate a cooperative spirit among members of the editorial department and to serve impartially the interest of society.
7. As far as the editorial side of the newspaper is concerned, the Editor vis-a-vis the Government and the proprietor is to be left free in the discharge of his responsibility.
8. In the event of disagreement between the board of management and the Editor leading to the Editor's resignation or removal, the Editor shall be entitled to six month's pay, gratuity and pay in lieu of the leave due to him at the time.

Journalistic Idealism

The Idealist, on the other hand, postulates that (i) a news event is basically what happens in the minds of the audience member, (ii) a news story is a device which triggers a perception or image in the audience member's mind and (iii) there is no 'news' except that which is perceived by the audience member. For the idealist, news is not news until it is reported, that is, perceived by an audience member through report.

1. The journalist as one is not simply writing for the education of others. He is also writing for self expression and he puts himself and his ideas into his journalism.

What he communicates is in a very real way what he himself is.

2. The journalist who has concern for ethics does care about or right action.
3. Such a concern indicates an attitude which embraces both freedom and personal responsibility. It leads to the discroses norms for action or as guiding principles or specific directives in propagating the kind of life which he thinks socially meaningful and satisfying.
4. Characteristics of journalists today, according to Merrill, a lack of commitment and consistency. Before any journalist chooses any particular ethics, he has certain social considerations in his mind.
5. Ethics is a branch of philosophy that helps journalists determine what is right in journalism. It is very much normative science of conduct primarily self determined or voluntary. Ethics is like self legislation and self enforcement.
6. Like their audience, journalists bring their own set of pre-conceptions in their writings and reporting pre concepti-ons produced by their environment and their position in the community. Their business relations also play some part.

The remedy lies through a right comprehension of the problem discussed and through a combination of conscious effort, changed perspective and new institutional arrangement with a view to eliminate systematic distortion.

Chapter – 4

Readability and E Journalism

READABILITY MEANING, DEFINITION AND FORMULA FOR READABILITY

- Readability means transmitting the connation into words and sentences that the average reader would understand and enjoy Crump (1974)
- Readability describes the ease with which a document can be read Klare (1976)
- Readability referred to making understandable by writing short words and short sentences and making it interesting by using many personal references Murphy (1977)

Readability Test

- Readability tests, which are mathematical formulas, were designed to assess the suitability of books for student at particular grade levels or ages.
- Readability formulas were first developed in the 1920s in the United States.
- Readability formulas today.

Today readability formulae are used to measure semantic factor (the difficulty of words) and one syntactic factor (the difficulty of sentences).

Readability research

Readability research, in general, takes into consideration several factors, such as average sentence length in words, percentage of simple sentences percentage of strong verb forms, portion of familiar words, portion of abstract words. Percentage of personal references and percentage two popular methods to measure readability are presented in brief *Reading Ease Formula* (Flesch, 1960).

The application of this formula requires the researcher to select systematically 100 words from the text, determine the total number of syllables in those words *(wl),* determine the average number of words per sentence *(si),* and compute the following equation.

$$\text{Reading ease} = 206.835 - (0{,}846\ wl + 1.015\ si)$$

The Reading Ease Score will put the piece of writing on a scale between 0 (practically unreadable) and 100 (easy for any literate person). The scale is as follows:

Description of style	Reading ease score
Very easy	90-100
Easy	80-90
Fairly easy	70-80
Standard	60-70
Fairly difficult	30-50
Very difficult	0-20

This is a measure of clarity of writing. A lower value of the index indicates easy reading, and a higher value difficult reading. To compute the Fog Index, researchers must systematically select samples of 100 words each, determine the mean sentence length by dividing the number of words by the number of sentences, count the number of words with three or more syllables, add the mean sentence length to the number of words with three or more syllables, and multiply this sum by 0.4. The graded Fog Index, against which the computed values are to be tested is as follows:

Performance of readability test

Readability tests can be performed manually by counting and doing a mathematical calculation, or by referring to a chart or graph. Computer can perform readability test. Things the readability tests can do

- They can serve as an early warning system to let the writer know that the writing is too dense. They can give a quick, on the spot assessment.
- They have been described as "screening devices" eliminate dense drafts and give rise to revisions or substitutions.

- In some organizational settings, readability tests are considered useful to show measurable improvement in written documents. They provide a quantifiable measure of improvement or simplification.

Things they can't tell you and why

- How complex the ideas are
- Whether or not the content is in a logical order
- Whether the vocabulary is appropriate for the audience
- Whether there is a gender, class or cultural bias
- Whether the design is attractive and helps or hinders the reader
- Whether the material appears in a form and type style that is easy or hard to read.

Because the readability formulae are based on measuring words and sentences, they cannot take into account the variety of resources available to different readers. Reader resources are word recognition skills, interest in the subject and prior knowledge of the topic.

Readability testing tools those are readily available Today most grammar software programmes provide more than one readability measures as well as comparisons to well known writing.

Readability Testing tools

- Word Perfect
- Microsoft Word
- EDIS authority tool
- Style Writer

How to increase Readability

Flesch (1946-60) has 25 rules of effective writing. They are

1. Write about people, things and facts
2. Write as you talk
3. Use contractions
4. Use the first person
5. Quote what was said
6. Put yourself in the reader's place

7. Don't hut the reader's feeling
8. Forestall misunderstanding
9. Don't be too brief
10. Plan a beginning, middle and end
11. Go from the rule to the exception, from the familiar to the new
12. Use short names and abbreviations
13. Use pronouns rather than repeating nouns.
14. Use verbs rather than nouns
15. Use the active voice and a personal subject
16. Specify, use illustrations, cases, examples
17. Start a new sentence for each idea
18. Keep your sentences short
19. Keep your paragraphs short
20. Use direct questions
21. Underline for emphasis
22. Use parenthesis for causal mention
23. Make your writing interesting to look at

Readability guidelines for authors of extension publications

Organization

1. Use headings and subheading to organize your document logically, if you use sub-heading, make sure to balance them by placing at least two sub-heading under any one heading.
2. Give your readers only the most important information

Writing style

1. Define unfamiliar or technical words and/or explain them with examples. Then repeat using them to make them more familiar to the reader.
2. Use shorter word. For example, use "do" instead of "accomplish"
3. Substitute compact words for more wordy phrases. For example, use "many" instead of "large number of"

4. Use shorter paragraphs
5. Use the active voice, not passive voice
6. Use positive statements. For example, "Practice safe pesticide application" is clearer than "Don't practice unsafe pesticide application
7. Use simple and definite example to explain concepts.
8. Use shorter sentences along with longer sentences.

NEWSPAPER MAKEUP

"Newspaper makeup" is defined as the design of a newspaper page the manner in which pictures, headlines and news stories are arranged on a page. The objectives of newspaper makeup are follows: To indicate the importance of the news to make the page easy to read, to make the page attractive

FRONT PAGE FOCAL POINT

Each page of a newspaper has a focal point on the page to which the reader normally looks for the most important story. Any area can be the final point, depending on the chosen design. Advertisements can also dictate the focal points of the inside pages of a newspaper.

On the front page of some daily newspapers, the focal point is often in the upper right-hand corner a now-dated practice that reflects the style of a bygone era. Americans, although trained to read from left to right and top to bottom, greatly altered this pattern for many years with respect to their newspaper reading habits. Through the use of banner headlines that extended more than half the width of the page, readers were trained to seek the upper right-hand corner of the front page, Newspaper readers begin their reading by following the banner headline across the page and continuing down the right-hand side of the page. Therefore, many newspaper readers have come to expect the most important story in each issue to or touch in the upper right-hand corner of the front page.

The right hand focal point not as important to makeup editors as in the past, since fewer newspapers use banner headlines on the lead story. However, many newspapers still carry the most important story in the upper right hand corner of the front page because of established practices.

Today, a large percentage of newspaper editors use the upper left hand corner as the focal point. These editors think that readers, trained in school to read other literature from left to right, prefer their newspapers to be designed that way too. A few editors still use other areas, such as the upper center of the front page as the focal point. Only time will tell which is best, if indeed, there is a "best."

INSIDE PAGE FOCAL POINT

The focal point on inside pages is the upper left-hand corner if there are no advertisements. Therefore, the focal point is influenced by a newspaper reader's natural sight tendencies and is not hampered by customs. On inside pages with advertising, the way ads are placed on the page influences the position of the focal point. The focal point is always opposite the lower corner of the page that is anchored by the largest mass of advertising.

MAKEUP LINES VERTICAL LINE

The vertical line is used to get the reader to read up and down the page. The line is carried out on the page by displaying stories, headlines and pictures vertically on the page. It is characteristic of the makeup of newspapers in early America and is still used to a limited degree in making up newspapers today.

Diagonal Line

The diagonal line is used in newspaper makeup to get the reader to read through the page. The line is carried out on the page by displaying headlines and pictures so together they forma diagonal line from the upper left-hand corner to the lower right-hand corner of the page. Also, a page can contain a double diagonal by forming another diagonal in the opposite direction from the first. The diagonal line lends a sense of rhythm to the page. It is characteristic of many of today's newspapers.

Circular Line

The circular line is used in newspaper makeup in *an attempt to get the reader to read* around *the* page. The line is carried out on the page by displaying stories, headlines and pictures on the page so the reader see each as being equally important. This creates a tendency on the reader's part to read all the stories. The circular line is used to a limited degree in modem newspapers.

Horizontal Line

The horizontal line is used in newspaper makeup to get the reader to read back and forth on the page. The line is carried out by displaying stories, headlines and pictures horizontally on the page. The horizontal line is a post World War II development and it is probably the most striking change in the appearance of newspapers this century. It is a characteristic of many present-day newspapers.

NEWSPAPER DESIGN CONCEPTS

LEARNING OBJECTIVE: Detail the design concepts used in ship or station newspaper makeup. Successfully designing a newspaper page encompasses more than experimentation. It is actually a calculated art evidenced by the following five newspaper design concepts.

BALANCE

In the balance concept, the page designer (hereafter referred to as the editor, although it may be any member of the newspaper staff performing this function) tries to balance heads against heads, pictures against pictures, stories against stories and artwork against artwork. This balance, however, is a relative balance, and it is not measurable but is something gauged in the viewer's mind. Therefore, the editor has sense, rather than measure, the balance for a page. This perception is one developed by experience.

CONTRAST

In the contrast concept, the editor strives to separate display items on the page so each gets the attention it deserves. The editor uses type, space and colour to achieve contrast. For example, the editor can achieve contrast with type by using regular type with bold face type. Headlines also can be contrasted by using bold, black heads or by displaying roman type with italic type. The editor can achieve contrast with pictures by using verticals with horizontals, small column widths with large column widths or dark and light photographs. Further, the editor can achieve contrast through color by displaying black type.

RHYTHM

By the rhythm concept, the editor tries to get the reader to move from one element to another element on the page. Rhythm is achieved in newspaper makeup by staggering headlines, stories and pictures on the page.

UNITY

The unity concept of newspaper makeup is used to tie the page together; therefore, the page is not divided one, two or more sections. A page that lacks unity is called a paneled page. You can avoid paneled by crossing the column gutters (space between columns) with headlines and pictures in the middle areas of the page.

HARMONY

The harmony concept is used to give a newspaper a standard appearance from day to day. Harmony generally refers to typographic harmony. This means using one type face for body type and a contrasting type face for cutlines. Headlines should have the same typeface as the body type and may be varied by weight and the use of italics on occasion.

The following components help the reader identify a newspaper NAMEPLATE. The plate should be simple in design, attractive, *and* harmony with the character of the paper. Its type should either harmonize or contrast with the headline type. The name plate can combine type and artwork together, The artwork however, should not make the nameplate jumbled and hard to read. The nameplate be made to float on the page. Although a name plate that runs the entire width of the page can be made to float, a floating nameplate usually occupies two or three columns and is placed anywhere in the upper third of the page.

FLAGS

A flag of the newspaper is a display used by a newspaper to indicate section pages or special pages, such as editorial, sports and family pages. Just like name plates, a flag should not dominate its page and should appear above the fold. Flags can also be floated (*Note*: Some authorities maintain that a flag is the same as a nameplate and identify a section head as a "section logo." We *do* not.)

MASTHEAD

A masthead of the newspaper is often referred to, incorrectly, as a name plate. A masthead is a statement that should appear in every edition to give information about the publication. These regulations also require that the masthead contain the following information: Name of publication, Date of issue. Frequency of publication, Issue number, Subscription price (if applicable), Name and address of the publisher, Second class mailing imprint.

HEADLINES

Headlines, or simply heads, contribute to all five concepts of newspaper design-balance, contrast, rhythm, unity and harmony.

The headline for one story should be separated from that of another. Heads that appear side by side (called "Tombstones") could be read as one head and confuse the reader. Tombstoning also prevents each head from gaining its share of attention.

When headlines and pictures are used together, they should be placed so the reader is not confused by their positions. You should not place a picture between a headline and a story, because the reader might being the outline thinking it is the first paragraph of the story.

Heads of the same column width should not be placed lower on the page than a smaller one, or higher on the page than a larger one. This does not mean that the bottom of the page cannot contain a large multicolumn head. It only means that heads of the same width should decrease in point size as they descend the page.

Do not run stories out from under their heads. This creates a readability problem by confusing the reader about find and finish reading the rest of the story.

A story can be wrapped (to continue a story from one column to the next) under its main head, or lead to achieve variation. A story is always turned to the right from its main part. A turn running above the headline of the story could confuse the reader and cause the individual to abandon the item. A story requiring a "jump" or continuation, to another page should be to split in midsentence, never at the period of a paragraph.

PICTURES

Readability studies have shown that pictures are one of the most popular elements in a newspaper. For that lesson alone, important pictures should be large and positioned in a manner that maximizes their display.

Pictures to two column widths or more should be placed on a page so they stand or hang from something that gives them support, A picture can stand on a headline, another picture or the bottom of the page. A picture can hang from a headline, another picture or the top of the page. A picture of two column widths or more should not float in copy, but a one column wide picture or smaller can float in copy.

Pictures and headlines that are not related should be separated by more than a rule, if the possibility exists that, when placed together, they are humorous or in bad taste. Reader's eyes have a tendency to follow the line of sight of people in pictures. Therefore, if people in a picture look off the page, readers look straight ahead or into the page. To prevent the reader from doing this, this main subject in pictures should go toward the center of the page whenever possible. This reader tendency can be used to you advantage be used to guide the reader's Try to avoid running pictures on the horizontal fold of a newspaper, because the area along the fold becomes distorted once the newspaper has been folded.

Do not give a picture more display space than it deserves, especially a "mug shot" (portrait type, close-up photograph of an individual). Mug shots can float in copy, but it is best to float it within a sentence a paragraph. Mug shots should be accompanied by at least a name line for identification. By omitting the name line, the reader is forced to identify the individual in the picture.

"Thumbnails" also are used in making up newspaper pages. The term refers to half column mug shots. A thumbnail is best used when it looks into the story or directly out of the page. A name line, in most cases, should also be used with thumbnails.

WHITES, GRAYS AND BLACKS

A newspaper page is "made up of varying degrees of whites, grays and blacks. Some pages may contain other colors. A good editor strives for relative balance of colors on a page and will not let any color dominate the page. You will not have any problems with white pages, black pages or any other colored pages: your concern in staying away from gray pages.

There are many ways to relieve grayness, or gray out, which is created by large areas of body type. One way is to use multi column pictures to break up columns of type. Another way is to use thumbnail photographs.

Type can be used effectively to relieve grayness. To breakup gray areas in a long story, you can set selected paragraphs in bold face type, if used sparingly. Another method of breaking up long gray stories is to use boldface subheads set about two pints larger than your body type size. A third method of using type of break up grayness is to use bold face, all-cap lead-ins. This method is particularly effective in matter set in wider measures. In two-column matter, the first three to five words of the paragraph containing a lead-in- can be set in bold face and all caps, and in one-column matter, the first one to three words of the paragraph can be set in bold face and all caps.

The paragraphs to be set in any of these bold faced methods should be the paragraphs that introduce anew element into the story or ones that contain information of more than usual interest. Two paragraphs using the same boldfaced method should not be run side by side because they tend to cancel each other out. Note that the use of boldface type is not favored by the editors of contemporary newspaper as much as by the editors with traditional leanings (More about traditional and contemporary designs will be presented later in this chapter). "Modem" editors rely on the use of different design concepts to eliminate large gray areas on their pages and consequently have little use for bold face type, except possibly as subheads.

Other useful devices in breaking up grayness are initial letters (mentioned earlier in this chapter), kickers and hammer-heads and sandwiches.

A sandwich is a device for handling "reefers" It is a small, sideless box made with the same rule used for regular boxes. The reefer type in the sandwich should be set is bold face and not be indented.

RULES

Rules are commonly used typographic devices in newspaper makeup. Properly used, they separate unrelated items and unit related ones. The two types of rules used are - the column rule and the cutoff rule.

Column Rule

The column rule is a vertical, thin line that runs from the top to the bottom of a newspaper page, depending on the width of the items to be separated or united. A cutoff rule is used to separate unrelated items, such as boxes, photographs, multicolumn headlines and advertisements, from the rest of the page. A cut-off rule helps the reader's eye turn the corner from where a story ends in one column to where it begins in the next column, except when the story wraps from the bottom of a page. Then no cutoff rule is needed.

E- JOURNALISM - ONLINE JOURNALISM

No other age would well qualify to be an age of technology than the present one. Change has become the order of life. In the sphere of journalism too, which contributes and supports the shaping of an age through its information gathering and disseminating functions, technology has made heavy in roads. In fact what we see and how we see is being shaped more often by technology than by human effort.

Computer technology has changed the mass communication model and changed communication across time and space. The digital transmission of text, audio and video altered the traditional one-to-many communication model. The internet and formation of virtual communication by people around the world show that society is now ready for the evolution of global village.

Online journalism stands to dramatically alter the traditional role of the reporter and editor. It places far more power in the-hands of the user, allowing the reader to challenge the role of the publication as the gatekeeper of news and information. Digital journalism has not developed its own rules and procedures.

Online journalism opens up new ways of story telling, primarily through the technical components of the new medium. Online journalism also can provide outlets for non-traditional means of news and information. The internet enables every one who owns a computer to have his or her own printing press.

The Internet and specifically its graphic interface the World Wide Web is reaching a level of saturation and widespread adoption throughout the world. Specifically for journalism practiced online in the discipline of computer assisted reporting (CAR) and a specific kind of journalism: online journalism - we can now identify and theorize about the impacts the global system of networked computers has had on journalism.

History

On 17 May 1991, the Geneva based research institute CERN released the World Wide Web standard. In May 1992 Chicago Online, the first newspaper service on America Online, was launched by the Chicago Tribune in the United States (Carlson, 2001). As of April 2001 the database of U.S. based Editor & Publisher Interactive contained 12,878 records of online news media. These journalistic ventures are defined as: "All media with a Web presence, one can find associations, city guides, magazine, newspaper, news services and syndicates, radio and television Web sites in the new database" (E&P Media Links, 2001).

Online journalism was evolved through three stages

The emergence of online journalism has not been an abrupt event. It has been evolving through three stages.

(i) Those who were looking for just a presence on the net did not invest much and that got reflected in their sites. Online journalists mostly reformatted content from their mother ship.

(ii) Most of the new sites were created. The journalists created original content and augment it with such additives as hyperlinks, interactive features such as search engines, which seek out material on specific topics.

(iii) In the stage the original news content was designed specifically for the web as a new medium of communication.

Even the online journalist can make decisions on which media formats best tell a certain story (multimediality), has to allow room for options for the public to respond, interact or even customize certain stories (interactivity) and must consider ways to connect the story to other stories, archives, resources and so on through hyperlinks (hyper textuality).

Elements that do not apply to online publications includes

1. **Time span:** Events that coincide within the time frame of publication are more likely to pass through media gates. Digital journalism allows constant updates. So timing plays limited role in decisions.
2. **Clarity:** Events whose meaning is in doubt arc less likely to pass through media gates. Hence online journalism has always clarity in it.
3. **Consonance:** Events that are congruent with an expectation may be passing through media gates.
4. **Composition:** Because gate keepers look at the day's news in its entirety, some news items are selected merely because they contrast with others.

Key areas for scrutiny by online journalists

Still now digital journalism has not developed its own rules and procedures, simply because the craft is too young. News on online may be less accurate than its traditional print counterpart. The speed of the internet has eroded the key standard of accurately verifying the facts of a story before putting it before the public. But the source of the problem goes beyond the speed of distribution. Without, staff, news organizations has no chance of checking and verifying. Both staffing and training inadequacies are obstacles to online news accuracy.

Some key areas for scrutiny by online journalists are given below.

1. Staff sizes and workloads for online employees.
2. Establishment of ethics protocols specific to online news.
3. Regiments for fact checking in the rapid fire world of e-news.
4. Clear and consistent placement of corrections and clarifications,
5. Labelling of advertisements used on the same page as editorial material.
6. Manipulation of images.

Advantages and Limitations of online journalism

Online journalism has both advantages and limitations. This is due to the many characteristics and issues specific to the medium. Some of the advantages and limitations are discussed below.

1. Immediacy

One of the most significant advantages of online journalism is its immediacy. Prior to the development of online journalism radio was the most immediate medium, with news bulletins every thirty minutes or every hour. However, the online environment allows for much greater immediacy. Journalists in this medium have the unique ability to publish news events as they happen. They can then update the stories to include subsequent information easily and speedily at any time of the day or night. As soon as the information is uploaded to the web, it is available around the globe.

2. Updated information

Online journalism promises more and updated news content than its traditional counterparts, most of it free. Some of the websites are going beyond their print counterparts. They can then update the stories to include subsequent information easily and speedily at any time of the day or night. Newspapers follow a 24 hours cycle, so they do their internet editions. But since internet is such a live medium, new sites are experimenting to keep pace with news in some ways. For e.g. The Deccan Herald site runs a ticker line at the bottom of the screen that flashes the latest news, which main paper will give in the next day's edition.

3. Medium-specific characteristics

Other advantages of online journalism relate to other medium specific characteristics. The incorporation of multimedia elements makes news websites more informative and entertaining. Hyper-links, blogs and discussion forums allow web users a more hands on experience, allowing them to be participants in the news process (Millison, 2004). The archiving ability of news websites is also a great advantage as it provides an extensive research resource to the public.

4. Reflecting changes in society

Another advantage of online' journalism is that the medium is reflecting changes in society as well as technology. About one hundred million people access the world wide web and the number is ever increasing (De Wolk, 2001 : 175). As online information can be accessed at any lime it is a convenient way for people to stay up-to-date on local, national and world news.

5. Hyperlinks

Online journalism features the utility of hyperlinks which is useful in both compressing and extending information depending on the reader's choice. The electronic cousin of a footnote belongs to a way of presenting information called hypertext. These are highlighted words or phrases, which when clicked, provide more information related to the given word or phrase. Then a window filled with new text pops up on the screen. But the extra material does not appear unless the reader asks for it.

6. Archival facility

The online publications are a sort for value - added product. Some websites provide complete version of print publications alongwith archives that run back a year. Indian Express online is one of them. The times of India made available all the information contained in Education Times, along with the databases of Indian educational institutions on its site.

7. Interactive editions

Internet news is a dialogue with an identifiable audience, a global conversation on topics that can be broad, such as on a general news site. The news sites deliver more news and information. They promote one to respond by e-mail or e-mail news letters. There are online newspapers which have services like 'news by e-mail, where one has to indicate the subject and they will get news and articles on the subject by e-mail from the online newspaper.

8. Emergence of webzines

A new category of news providers, apart from newspapers and television channels, has also emerged on the internet. They are the websites, which do not have a print or TV counterpart, but run news service on the internet.

9. Easy skimming and scanning

An electronic newspaper facilitates hunting, browsing and or grazing. An 'autopilot mode' would flash headlines at a speed set by the reader. If something looked interesting, the readers could press a button to stop the flow and read the entire story. At any point, the reader could look something up a difficult word, a map of a country, a story from last week and or last year.

10. Personalised context and atomization

Cyber news allows news consumers to understand the meaning of the day's events in a personalised context that makes better sense to them than traditional media do now. Since it can build new communities based on shared interests and concerns; and since it has the most unlimited space to offer levels of reportorial depth, texture, and context that are impossible in any other medium — new media can transform journalism.

11. Job opportunities

For the news professionals, online journalism has created more job avenues. A new genre of specialised personnel is emerging in the form of web editors, content providers, web visualizers, web graphic designers, etc.

12. Revenue

Most of the newspaper sites in India and abroad are free. In some cases, a few services like archives and other value added products arc charged. But high subscription magazines like the Economist and scientific journals like Science and Lancet are open only to subscribers of their print editions. The experience of far shows that some of them have become self sustaining in the short period they gave been on the net.

Limitations of online journalism

1. Despite the numerous advantages of online journalism, there are also limitations. The benefit of immediacy can give rise to some serious ethical issues. The desire to publish brand new information and the ease of which it can be altered may cause information to be made accessible before it is verified. This undermines the journalistic principle of accuracy and can lead to misinformation.
2. Other limitations are more general. The immense size of cyberspace and the extent of information available may

intimidate audiences and cause information overload (Hall, 2001). Furthermore, it may be difficult for audiences to distinguish between credible news websites and other non-official news websites. This can lead to confusion and misinformation.

3. There are also limitations in online journalism regarding the issues of surveillance, censorship and privacy

Applications of Online Journalism

- Characteristics
- Modern Applications
- Ethical Issues
- Future

Online journals in agriculture, agricultural extension and related fields

Online journals provide a variety of information. They can be accessed at any time from anywhere provided there is a computer with internet connection. Some of the online journals are listed here.

1. Journal of Extension; www.joe.org
2. International Review of Research in open and Distance Learning:www.irrodl.org
3. e-learn magazine: www.elearnmag.org
4. Journal of Interactive Media in Education: www.jime.open.ac.uk
5. The Electronic Journal on Virtual Culture
6. The Interpersonal Computing Technology Journal 1. (IPCT-J)
7. The Distance education online Journ 41
8. Sociological Research online
9. Journal of sustainable agriculture
10. Australian Journal of Agricultural Research
11. British Journal of Educational Technology
12. Journal of Nature Conservation
13. Social Science Research

Chapter – 5

Forms and Techniques of Writing

TYPES OF INTERVIEWS

Art of interviewing

The work of the newspaper reporter is, in a sense, one long interview, for he obtains the majority of his information by asking, people question and noting the answers, but the published interview in which a writer gave word for word answer and described even the actions of the interviewee is no longer in fashion. This technique is employed only in reporting press interviews of film stars or cultural dignitaries.

Manner of Introduction to the Personality

An interview begins at the moment you sending your card, or knock on the door, but the success or failure of it will depend entirely on the impression you make on your entrance. It is at times like this that the affinity between the press and the public personalities becomes really close, for the journalist must often be something of an attraction for the publicity Hungary person.

To begin with, so much depends on how you pass the time of day. Have you ever noticed, even in general life, what effect your 'good morning' or 'good afternoon', can have upon people, with the next three or four acquaintances you meet, try greeting them in various ways as an experiment, and notice the different reactions. The responses obtained will be interesting, and for the junior who cares to observe them they hold important lessons. There are, indeed several ways in which a 'good morning' can be said. The surly grumpy person spits out the words as if getting rid of an unpleasant taste the preoccupied says it automatically and in a monotone, the alert will utter the words in three tones; he will say 'good' as if he really means it, 'morn' on a slightly lower pitch, and the 'ing" on an upward inflection so that it finishes a tone above the 'good'. And when it is said in that fashion, and with a smile, notice the reaction. Almost right away the person spoken seems to brighten up, and he gives a more ready response than if the word 'morning' was spoken as if it came from the bottom of a deep pit. He says to himself : "Now here is someone who is not merely saying good morning' because it is customary. He means what he is saying; he obviously enjoys his 'good morning' and the fact that he is wishing me one is a compliment!" Immediately he is m a receptive mood, and even if he intended to refuse an interview, the chances are that he will give up to you after all, if you play the rest of your cards correctly.

Give the Lead and then listen

It is a good and wise plan to decide not only on your opening, but also on the first three or four points you wish to raise. Much depends on the nature of the interview. It is to obtain views on something that has happened and the person being questioned has had "no time to consider the situation, try and give him a lead. In order to give him a few moments, tell him, as sincerely as you can, that you have asked him for his views because they may be of assistance to others. Stress the usefulness to the public generally of knowing the ideas to make this piece of flattery too obvious, but say it in a way that will convey. Sincerity throughout the interview, do not more talking than is absolutely necessary'. It is not unknown for reporters, especially when a little self-conscious, to dominate the interview almost *to* the point of boredom instead of giving other man his tongue and only making a comment again. The best reporter is he who can listen.

Manner of closing

It is advisable, before departing, to spend. A minute or two discussing commonplace matters. A quickly ended interview, *in* which the reporter puts away his paper, stands up and bids good day creates a bad impression. It leaves the person interviewed with a feeling that, having pricked his brains, you have no further use for him. Your last words should be: 'Well, goodbye and thank you.

The written interview where a person writes answers to a series of questions sent to him is now seldom employed. Perhaps this is as well, for it is never wholly satisfactory. For one thing, a 'follow up' question can never be asked and so happen that a answer written almost demands further questions in order to elucidate a point. In addition, it lacks the personal touch that essential. Avoid it as much as you can.

Types of interviews

The Telephonic Interview

In recent years many papers have adopted the technique of the telephone interview; a method which has the great advantage of saving time for the reporter; but that is all. It is a selfish way of doing about things, for not only the person again lacking, but it assumes that the paper is more important than the person being interviewed and that is a thing the person being interviewed should never be allowed to fee). It is possible that he may unnecessarily be on his guard, for any one with an ulterior motive could ring him up and profess to be a newspaper representative; on the other hand, he many for some reason dislike using a telephone, which immediately places you at a disadvantage.

It is willingly conceded that on evening papers especially a telephone interview is sometimes necessary, but in general, it is a lazy and bad-mannered way, particularly with a stranger. With someone to whom you happen to be well known the telephone may occasionally be a useful way of obtaining your information, but taken by and large, it should only be used in the case of emergency.

The telephone interview has been found as a useful tool of getting the news of local people when you are very near press time, which perhaps accounts for the extension in its use. For instance, suppose some restriction on a commodity was suddenly lifted, and you were told to get interviews for a local angle story: you would need to contact the President or Secretary of the Association concerned, one or two leading people, and a housewife. From them you should obtain a pretty good variety of views, some of them welcoming it, others fearing that the sudden lifting of the regulations would cause such a run on it that the commodity would be virtually unobtainable, and so on.

Before ringing up these people, try and put yourself in their position: if you were contacted by the ringing of a bell and asked for your views, would you be able to give them, on the spur of the moment? Even if you could, would you be prepared to do so, knowing that whatever you said would be made public in the four corners of your locality? Would you not prefer to have a few minutes in which to give a little thought to the matter? These are points which the thoughtful reporter will bear in mind, and it may well be, as a result, that he will first ring up his people and ask them if they would give their views when he rings up again in ten minutes' time. Many people are only too willing to assist, but they like to have a brief period for reflection before committing themselves and they appreciate the friendly tip that you want them to do so.

When has been said above applies to all reporting work and not merely interviewing? Put yourself in the other fellow's place; try and understand his position. By doing this you will be helping yourself immeasurably; you will be cultivating a completely detached and unbiased outlook. Thus you will be better able to appreciate that there is another side to every question, and because of that you will gain variety of approach when hunting for news. Many young reporters today take so much for granted that they become their won worst enemies, simply because they either refuse to Do's:

1. As far as you can, study your man and the subjects which you wish him to talk.
2. Show interest in the job and try to get easy terms with your man quickly.
3. Go on the job well dressed but not over dressed.
4. Get an understanding at the outset that you are at liberty to publish any of his remarks unless he asks for special points to be regarded as confidential and as 'off the record'.
5. Submit a copy of your interview to the person interviewed if time permits and see that when he returns it, he corrected every page.
6. Always remember that there is such a thing as of libel and that the indiscreet journalist personal remark may cause a great deal of trouble. Don'ts
7. Don't think that you "know more about the subject than the man you are interviewing, because you definitely do not.

8. Don't adopt a servile attitude, but at the same time, show-that you respect your man on what he has done.
9. Don't outstay your welcome. When you feel that you have got all you require do not hang about talking of crops or cricket. Remember, that he may have another appointment even if you have/not
10. Don't feel hat it is necessary to argue with your man if you do not agree with some of the views he expresses, because after all the interview is not at all a debate.
11. Don't flourish a notebook in your mains face, although it was the newspaperman's Bible. To appreciate, or incapable of appreciating that there are more points of view than their own.

Specialised Interview

In the case of important individual Interviews only a senior reporter generally given the assignment well in advance. He spends few days in studying the career of the great personality to be interviewed and frames the questions to which he wants an answer. If the interview is to deal with a specific subject and is not intended to be a general survey of the world or of the locality in which the newspaper is published, the reporter must make certain that he has at least a rudimentary idea of the matters which the wishes his personality to talk about. It is no good being ushered into the great man's presence and then waiting for something to turn up.

The reporter would be well advised to utilise the resource fulness of the office librarian who can generally provide a host of ideas and cuttings, when he is told the name of the man to be interviewed.

The reporter is sometimes asked by the personality to submit the receipt of the interview report and if time permits, he should comply with the request and make the stipulation that when it is returned, the words O.K. should appear on every sheet. There can then be no post-mortem either on the interview or the reporter. But it must be made clear that the general sense of the interview must not be altered and that in agreeing to the request for a script the desire is that the interview should be rounded off and that occasional words should be altered if they do not entirely convey the intended meaning. There are a few 'Do's and 'Don'ts' which journalists should keep in mind while they are functioning as reporters.

Press conferences

There are frequently occasions for Press Conferences at which ministers or leading figures in the world of industry, science, entertainment or sport wish to inform the newspapers of latest developments. In such cases, the Editor gives instructions to the Public Relations Officer to get the journalists together and hear the pronouncements if any. It is generally left for the PRO or the News Editor to decide whether he will make it a 'free for all' conference or a gathering limited to specialists like the Parliamentary or other Correspondents. Generally, a Minister prefers to meet a party of experts because he realizes that knowledgeable and sensible questions will be put to him after his pronouncement. There are sometimes Press Conferences and interviews which are simply a waste of time, where the information could have been issued by the Public Relations Officer of the Ministry concerned in the form of a 'handout'.

Question-Answer Formal Interview

Another type of interview which has been adopted in a country like the USSR is that from time to time an enterprising foreign correspondent submits a series of questions to the Soviet, chiefs. If he is fortunate he receives in due course a series of answers to his questions and these are usually handed over at the same time to the official Russian news agencies for general circulation. But they are written answers and thus there is no chance of obtaining the realistic atmosphere which is the way of the cut and thrust of personal cross examination. It is quite certain that this type of interview would not inspire any ambitious reporter with great enthusiasm if he were enthusiastic with such an assignment.

ELEMENTS OF NEWS / QUALITIES OF NEWS

There is, in fact, no scientific formula for recognizing or judging the value of news, precisely because news - the gathering of it, the writing of it, and the reading of it contains a "Human" characteristics. It is perilously subject to the whims, fancies, and caprices of the human mind. Working with such human dimensions might better be described as an exotic mixture of witchcraft, poker, and psychoanalysis.

Reporters are not totally without criteria or elements / qualities by which to judge news, however. Basically the criteria listed below are rough gauges of what interests the people, or what should interest them if they are to be useful citizens of their community. While these criteria are not precise, they make guessing a little more systematic.

The basic elements / qualities / criteria of news

The ten criteria listed here are designed more for the person gathering news than for those editing news or laying out the pages of a newspaper. They represent the essence of what a reporter should look for in talking with a news source.

1. Timeliness
2. Imminence
3. Consequence
4. Proximity
5. Change
6. Action
7. Concreteness
8. Personality
9. Rarity
10. Conflict

1. Timeliness

Something that happened last night is more newsworthy than something occurring three weeks ago. News must be "new". When you read or hear it, it should be fore the first time. New revelations of old, therefore, unknown events can be news, however, as for example, the discovery of a sunken eighteenth - century Spanish galleon in the harbor of your community. The discovery is the news, not the sinking 200 years before. Some newspapers insist on getting the word "today" in every news lead no matter how old the event, sometimes with ludicrous results: "Firemen today are still talking about the tenement fire that occurred two weeks ago yesterday". Still the reporter always seeks the latest development even if there is little to report : "Police today announced discovery of a bloodstained hand-kerchief in their continuing investigation of last week's brutal slaying of a teenaged cheerleader". In agriculture seasonal related issues are simply news.

2. Prominence

More news value is placed on better-known persons *or* institutions in the news. People are simply more interested in kings and celebrities than in peasants and average citizens. A minor auto collision is a bigger news story if the governor is one of the drivers, bigger till if the President is riding in the car. The same is true with institutions. The announcement that Harvard University is expelling two dozen students caught at cheating has more value than if Ponderosa College does so. Prominence comes by 1 power, II by position and III by Elderness.

3. Consequence

The impact of a news event on the public is one of the key elements in judging news values. A story of tax reform affects every taxpayer in the state; therefore it has high news value. If the governor is killed in an auto accident, the repercussions are widespread. His removal from the political scene may bring significant changes in the nature of state government, and these changes affect all citizens. Often a reporter must seek to learn the consequences of any act, no matter how seemingly small. If a lot of false fire alarms are being turned in, what are the consequences of that? They may be dire indeed if fire equipment is out on a false run when a real fire occurs.

4. Proximity (Nearness)

The nearer the event, the greater its news value. Although some news has national importance activities of congress or the white House, for example - most community newspapers thrive on news close to home. An escaped convict in Texas means nothing to a reader in Minnesota - unless the convict is from Minnesota or is known to be headed in that direction. On balance, tomorrow's forecast for balmy weather in your community has more news value than a hurricane threat 5,000 miles away, simply because it makes a difference to you. You may go to the beach tomorrow. The size of the community also makes a difference. A bloody but routine murder will make front page headlines in a small community, but it would hardly rate more than a paragraph on page 46 in Chicago or Los Angeles. Proximity means not only physically but also emotionally, culturally, professorially etc.

5. Change

If the reporter can point out that a certain condition or activity represents change, then the item carries more news value. People pay more attention when it is clear that things are not the same anymore. The change might be more potential than actual, but it remains no less important. People may be cheered or threatened by impending change, but they want to know about it: may be they can even do something about it. Sometimes lack of change can be news, such as increasing crime rates, higher cost of living, or increased unemployment. Sometimes it takes an alert reporter to spot trends: women are wearing their hair shorter this season, or more and more men are growing beards. Sometimes it takes an unfortunate incident to call attention to change, increasing traffic hazards near a school are evident only after a schoolchild is run over and killed.

6. Action

The concept of people doing things is always more dramatic and therefore more newsworthy than people merely thinking or expressing opinions. Take the problem of increasing traffic hazards, for example. The fact that parents are vaguely dissatisfied with the traffic control near the school is at best marginal in news value. It's more interesting if they take some king of action, submit petitions to the police chief, picket city, hall flock to the city council meeting to register their protests or file a suit in court. If there is more of dramatic or violence in action greater is the value news. The most extreme would be physical violence, murder, assault, or mass bloodletting ranging from riots to war.

7. Concreteness

The tangible always takes over the abstract on the scale-oil-news values. That's why the existence of the dead child - killed on the busy highway near the school has more public impact than vague commentary about how dangerous that street may be to children. Similarly, the reporter who "shows" (by concrete and specific examples) will have stories of higher news potential than the reporter who merely "tells". The greater the amount of concreteness is a story, the higher the news value". That's why the $ 100 robbery.

8. Personality or "human interest"

People are naturally curious about other people, particularly those who are celebrated or somehow "different". In large measure this may be because readers identify, positively or negatively, with such persons and with any interpersonal relationships that they may conduct (friendships, rivalries, romances, sex liaisons, partnerships, and so forth). Quirks, idiosyncrasies, fantasies, secret desires all of these attract high readership, particularly if the person involved is prominent.

9. Rarity

Like a precious metal the more infrequently a certain kind of incident occurs, the more news value it contains. This is the classic "man bites dog" a high - wire artist thrills thousands with his exploits but earns scant attention from news media, then winds up on page one when he falls out of bed and breaks his toe. It is a simple human characteristic that if you see a 100 member marching band with 99 wearing green uniforms and one in yellow, your eye is caught by the yellow. You are curious. Why is he different?

10. Conflict

Sometimes the conflict is inherent in the situation, as for example, a confrontation between college students and police. Just as often, however, it is the reporter who recognizes that at least two sides exist in more situations, and he seeks to present the other side. If for example, the district attorney remarks in a press conference that the community's judges are too lenient in sentencing convicted criminals, the reporter would call several judges for their reactions and would incorporate their comments into the story. The existence of conflict or controversy implies that an issue remains unsettled, thus it is interesting and significant because it contains both consequence and the threat/promise of change.

Ex: Sport news is conflict type of news. Ground water depletion due to Eucalyptus plantation is a debatable issue and is a conflict in nature.

The reporter typically finds few news events containing only one of the above elements criteria. Mostly a combination occurs: a prominent personality in your community takes a rare but violently concrete action the consequence of which is timely change. That should give a reporter something to write about.

Additional elements / criteria

A good reporter learns how to make his stories sparkle, partly through the use of clean, crisp prose and partly through the use of one or more of the following criteria. These criteria, while not part of a basic list, can add to the interest and understanding of a news event or situation. Sometimes the use of them depends less on the inherent nature of the news event and more on the reporter's "literary" skill in gathering and presenting information.

1. Celebration
2. Adventure
3. Mystery - suspense
4. Drama
5. Ethics
6. Self- improvement

1. Celebration

Much news is actually of the celebration category, much as you observe a birthday or an anniversary. You age a little every day but you don't acknowledge that fact until you pass a milestone such as a birthday. So it is with news" "Today marks the 10 anniversary of the great flood of 72 that took a dozen lives and caused $2 million damage". Celebrations may also come in the form of civic parades, ribbon cutting ceremonies, speeches proclamations, publication of annual reports awards dinners, retirement parties, and weddings.

2. Adventure

Readers get a vicarious thrill in learning of the physical or intellectual exploits of daring men and women. Thus a daring climb of a difficult mountain or a 40-hour solo flight across the pacific becomes newsworthy, but subtle adventures belong in the news, too, such as a policeman's emergency delivery of a baby in a car or a swimmer's rescue of a drowning child from a turbulent river. Such stories often can be presented in narrative form that heightens the suspense. The adventure can also be intellectual, such as an account of a political debate or the exploits of a scientist seeking to unravel the mystery of the Bermuda Triangle.

3. Mystery - suspense

Enigmas intrigue, people. Sometimes the mystery is implicit in the news event, what happened to the missing five year old girls who disappeared from her home two months ago and hasn't been seen since?" does life exist on Mars? If not, how about other planets?

4. Drama

Sometimes factual situations can be presented with acting and character insight of a short story, complete with narration dialogue, and plot. The classic plot pattern of much fiction writing is worth study by news reporters. It starts with a character meeting a conflict. The character struggles with the problem through a series of events meet a final crisis. He finally conquers it (happy ending) or it conquers him (tragedy). Many situations are inherently dramatic because they involve one of the three basic types of dramatic conflict: man versus man, man versus nature, man versus self. Obvious drama is present when a pilot brings a crippled airliner, trailing black smoke, into the airport for a tense but safe landing. Other stories could be of a more subtle man versus self struggle. One could imagine a journalist turning out, through skilled interviewing and sensitive writing, a first - rate literary study depicting the inner conflicts of, say, a convicted murderer.

5. Ethics

With the history of human endeavor being a constant search for what is right, reporter finds interest and news value in this area, too. Possible news articles can range from unethical practices of government officials to examples of injustices perpetrated on he "little guy" by a heartless bureaucracy.

This category, though it may stretch the traditional definition of news a little, serves mention. Readership studies have suggested that stories with high utility value have high interest. These include stories of home maintenance, food finances, human relationships, entertainment, travel health, beauty, and fashions. Whether they are news in a traditional sense depends largely on whether they are new - new ideas, new trends new reasons to accept old explanations, new advances in nutrition, medicine cosmetology, and so forth.

WRITING NEWS STORIES AND INVERTED PYRAMID

The newspaper story in modern society has evolved to meet the requirements of every day life as lived by everyday readers. It relies on the elements of novelty, directness, pace and variety, and it strives to convey its information in the form of most in keeping with the tempo of our times. It aims to state its facts quickly and clearly. What is important in News Stories?

While writing a news story the reporter's worry is to give it a proper lead which would attract the reader and keep him attracted till the end. There are many kinds of leads and the reporter has to choose the most appropriate one for his story. The different types of lead are Staccato lead, Explosive lead, Dialogue lead, Punch lead, Asterish lead, Contrast lead, Descriptive lead, Epigram lead etc.

In writing a news story the style is important. A well known journalist has written: "The language is perceptually in a flux, it is a living stream, shifting, changing, receiving new strength from a thousand tributaries, losing old forms in the backwaters of time.

Parts of News Stories

A news story has three parts: viz., (i) The headline, (ii) The first paragraph and (iii) the remainder of the story.

The head line first attracts us: It stands out in bold black type. Its message is terse, abrupt and often startling. It makes us stop and look. It tells us quickly what the story covers. Its primary function is to attract our attention. Headline writing belongs to the copy reader's province and not to the reporters.

The first paragraph or lead: It would be difficult to overstate the importance of the opening paragraph or lead. Always this lead remains the primary concern of the newspaper writer. In any form of writing the writer tries to put his best food foremost. In journalistic writing, this first stride has unique importance.

The remainder of the story: From the headline and the lead we come to the rest of the story. The reporter constructs the model news story after this pattern. He selects the most important incident or fact for his lead. Then he proceeds by selecting the next important incident, fact or detail, then the next important, the next most important after that and so on till he reaches the least important phase of all. Guided by his idea of news importance, the story assumes graphically the shape of an inverted pyramid.

Points to be noted in writing a News story

1. It should be compact, using short sentences, with every word selected and placed for maximum effect.
2. Short tense paragraphs each complete in itself and generally capable of being removed without destroying the sense of the story.
3. Conciseness, directness and simplicity through elimination of superfluous words, phrases, clauses and through proper emphasis.
4. Be factual keeping out editorial opinion wise superlatives, adjectives, nouns or dogmatic words
5. Avoidance of fine writing, strong verbs and nouns being preferred.
6. Observance of the rule of good grammatical and word usage.

Structure of a News story

A News story can be separated into five parts.

1. An "opening" designed to capture the essence of a situation clearly and dramatically. The opening or lead need not be confined to one paragraph. The first sentence of the story, however, should be carefully calculated to draw the reader into the second sentence and so on through the story.
2. A "Catchall paragraph" that elaborates on the lead and provides any necessary details. The lead and the catchall, working together, should answer the who, what,

when and where, leaving perhaps the why and the how for the subsequent paragraphs.

3. A "sales message" if appropriate this could be an appealing quote or statement of what an event or situation means to the story reader.

4. A transition to the details.

5. "Details" as needed to complete the story. Generally the details should be organized into blocks with each block representing a unified set of details. The more complex the story, the more you have to work to confine material into blocks, like chapters in a book. Again, these blocks are usually organized in order of descending importance.

Lead tells the story content

If interested, the reader can continue. Copy reader can easily cut the story from the bottom and yet retain its readability. If the bottom of a news story written in the inverted format were cut, the top should still provide enough facts to give the reader pertinent information.

TYPES OF LEADS

The first paragraph or lead: It would be difficult to over state the importance of the opening paragraph or lead. Always this lead remains the primary concern for the newspaper writer. In any form of writing the writer tries to put his best food foremost. In journalistic writing, this first stride has unique importance.

Main Types of Leads

(a) The digest lead: This lead summarizes clearly all the principal facts. It remains the simplest in construction and forms a basic part of all lead variations.

(b) The direct appeal lead: This form borrows the interest compelling device of the personal letter. It addresses the reader directly or by implication as "you", and has the effect of making the reader a collaborator in what follows. It often begins with such phrases as "If you have ever thought", or "If you have ever seen or read". Here is a typical direct appeal lead.

(c) The circumstantial lead: Here we have a beginning which stresses the circumstances under which the story happened. It crops up usefully when the story has a human interest slant.

(d) The statement or quotation lead: This type of lead starts out with an enunciation which as often occurs in quotation marks. In speech reporting particularly, a succinct, epigrammatic sentence often puts in capsule form the gist of the speaker's idea. Where the reporter has added to this the inevitable five W's, the resulting paragraph coincides with the digest or summary lead.

(e) The Descriptive Lead: This form of beginning presents a picture. The reporter may set the stage for the action of his story or he may present in detail one or more of his chief actors. In other words, this lead can describe the scene or it can describe one or more of the people involved.

(f) The suspended interest lead: This type of lead serves as a stimulator of interest. It gives the reader enough information. After the lead, the story usually runs along in chronological form, so that the reader must read to the very end to get at the climax. Reporters use this type of lead chiefly for short bits, on the theory that if used on longer articles, readers would not bother to wade through paragraph after paragraph. Here is a typical suspended interest lead, used in a typical way.

(g) The tabulated lead: Occasionally one runs into a story in which no one fact is prominent. Each facet of interest has about the same value. In such instances, a practice has grown up of tabulating each item in the lead - one, two, three, four.

(h) Various stunt leads: Accordingly a variety of what we can call exclamation point leads crop up from time to time and have been dubbed, "astonisher", "punch" and "teaser" as the case may be originally one knows no pigeonhole and refuses the standardization of definition beyond the generalization that such leads are eccentric variations from the norm. When successful, a stunt lead has its own reward.

(i) Single sentence lead

(j) Multiple sentence lead

(k) Quotation lead: The lead contains direct speech of the person with whom the story is related. The main purpose of this type of lead is to draw interest of the readers by quoting important personalities.

(l) Question Lead: In some situation, the lead is composed by putting some question inorder to highlight the problem situation and also to draw attention of the readers.

(m) Descriptive Lead: Its main purpose is to describe the news story in such a way that the readers imagine in their mind the place, time, milieu of the fact while going through the news. The excitement, emotion, happiness, sorrow etc. are all intermingled with the fact to touch the heart of the readers.

(n) Delayed Lead: Instead of presenting the news story, the first paragraph is preserved by suspense and excitement which is to enhance the readers' attention and interest. Readers then proceed to the, subsequent paragraphs to know the unknown. This is called delayed lead.

(o) Dialogue Lead: This type of lead is written with the exact conversation between two persons. Sometimes it may be between the reporter and an eminent person with whom the news story is related.

(p) Staccato Lead: Sentences are sequentially arranged in a rhythm which cherishes the main fact. It is just like a collage i.e. a picture made from scraps of paper. This type of lead is applicable in a tragedy which was unpredictable and unbelievable.

Example: Times News Network

Hyderabad: About 500 farmers died in Warangal district due to exposure to pesticides during the last cotton crop season and another 1,000 were exposed to related health hazards, according to a fact finding team that toured the district.

(q) Bullet Lead: It is composed with a short sentence. It will pierce into the minds of the readers like a bullet. The sentence is short, but strong enough to stir up the readers' mind. Generally, radio and television use such type of lead.

WRITING FEATRUE STORIES, STRUCTURE

According to Alexis' McKinney of Penvery Post (Rivers 1967), a newspaper feature story is an article which finds its impact outside or beyond the realm of the straight news story's basic and unvarnished who-what-where-when-why and how. As Rivers (1967) puts it, there is no standard form for features; the middle and the end should be as compelling as the beginning. He adds that the facts that a feature writer uses must be informed with colour and life, background and interpretation, an imaginative way of perceiving people and situations, and sharp bright phrases.

Feature, as per Hough (1998), is a term with a number of meanings.

(i) The main news angle or news peg of a story. A special point or twist to a story (News peg: the significant or interesting point on which a reporter writes a story)

(ii) A type of news story, sometimes called a news feature. Features are often

(iii) interpretative, give background, play up human interest and convey the colour of an event

(iv) Non-news mailer, for example, syndicated columns, cartoons, horoscopes and so on, that appears regularly in the newspaper

(v) To give something prominence in a story; to emphasize something.

According to Metzler (1979), if reporting is fun, then feature writing is the most fun of all. It's also the hardest. But if such a thing as "creative" writing exists in newspaper reporting - and I think it does - then surely it comes in that nebulous literary form known as the feature story.

Structure and organisation

The feature story offers considerable freedom to the writer, but a structural pattern does exist and should be learned. A structure particularly appropriate for features is the structure that has not only a lead and a development, but a carefully contrived conclusion.

Lead ----- Body of story ------Conclusion

The overall pattern of the feature, looks like this.

(i) An opening calculated to gain reader attention and interest also to indicate what the article is about. Often it is an illustrative anecdote or a descriptive scene to set the stage.

(ii) A general statement of theme in which you orient the reader specifically to what the article, is about and perhaps what means to the reader of how it ties in with some news event emerged. This "orientation" section may be brief or it may be explicit, taking several paragraphs.

(iii) Two or more main supporting points for the theme. These are like topic headings for an essay or speech. They comprise the body of the article.

(iv) Ending. Your ending should be strong. If your feature is well written, many readers will follow it all the way to the end. They deserve *a* reward, like dessert after a meal. Do it with a colourful anecdote, a touch of humour, a worthy quotation, or a food-for-thought comment. Many endings try to tie together with the beginning, so that a question raised in the opening paragraph is answered to the reader's satisfaction in the end.

Suspended - Interest Stories

A very specialized story form is the suspended - interest or pyramid, form. This format is used most frequently to toil short, humorous stories, the kind editors call page brighteners. The suspended - interest story is tricky to write because the writer must be careful, to save the best for last.

Structure of Suspended - Interest story

The suspended interest story is carefully constructed so that the lead does not give away the outcome of the story. The details are arranged and also carry the reader forward to the final paragraph where the outcome - the punch line - is revealed.

Feature Writing

The feature differs not fictional. Features differ from editorials, reviews, and most columns in their approach to opinion. It is the difference between making reasonable judgments and advocating. The feature writer often judges, he should never advocate, He aims not at persuading but there is no standard form for features; the middle and the end should be as compelling as the beginning, The feature writer must think of his writing as thematic, with the sentences and paragraphs tied to each other in such a way that a unified, flowing story is produced. Although many feature stories are quite as timely as news stories, the feature writer must not depend on timeliness for reader interest. The simple presentation of facts growing out of recent events is not for him. Instead, the facts that he uses must be informed with colour and life, background and interpretation, on imaginative way of perceiving people and situations, and sharp, bright phrases.

TYPES OF FEATURES INTRODUCTION

Feature articles are informative stories which rely on human interest and emotions for impact. They can often be published a week or a month later and still be relevant. Features provide more detailed background information about the person, situation or issue rather than focusing on the latest news development

Features are not meant to deliver the news first hand. They do contain elements of news, but their main function is to humanize, to add colour, to educate, to entertain, to illuminate. They often recap major news that was reported in a previous news cycle. Features often:

- Profile people who make the news
- Explain events that move or shape the news
- Analyze what is happening in the world, nation or community
- Teach an audience how to do something
- Suggest better ways to solve
- Examine trends
- Entertain.

A feature is not an essay, It is a news story that has been lengthened with more fact, more background and more interpretation by the reporter. Feature writing offers great satisfaction to the journalist, but also a great deal of hard work. Every fact and angle must be research. Feature articles should be planned step by step. The composing of the feature should not be rushed. The writer must ensure that the text is organised and thoroughly researched. A well written story will make readers want to read the whole article.

Types of Features

Personality profiles: A personality profile is written to bring an audience closer to a person in or out of the news. Interviews and observations, as well as creative writing, are used to paint a vivid picture of the person.

Human interest stories: A human interest story is written to show a subject's oddity or its practical, emotional, or entertainment values.

Trend stories: A trend story examines people, things or organizations that are having more impact on society. Trend stories are popular because people are excited to read or hear about the latest fads.

In-depth stories: Through extensive research and interviews, in-depth stories provide a detailed account well beyond a basic news story or feature.

Backgrounders: A backgrounder also called an analysis piece adds meaning to current issues in the news by explaining them further. These articles bring an audience up-to-date, explaining how this country, this organisation, this person happens to be where it is now.

Bright: Described as a "brite" is a tiny human interest factor, usually written in anecdotal style with the climax at the end (suspended interest).

Sidebar: This is a feature story that accompanies a main news story. A story about a major flood, for example, may include such sidebars as an interview with a refugee family, a background story on what caused the flood, or a colour piece on the work of rescue units perhaps even a narrative of a single dramatic rescue.

Organisation or project profile : Similar to the personality sketch or profile except that the organisation focused article deals with groups or agencies rather than individuals. An example would be a story about a group of women who form a committee to secure better treatment of battered wives and take their case to courts, legislators, police officials, and so on.

News feature: This is a news story written in feature style. Instead of a straight, unvarnished account, it will use feature techniques, such as opening with an illustrative anecdote, even though the story's major purpose is to convey news.

Comprehensive news feature: This kind of story depicts trends about the ongoing news/issues rather than scientific news events. It is better researched than most daily news stories since it comes from a wide variety of sources. It is usually more analytical and interpretative; it tries to depict not only what the news is but what it means.

Personal experience article: Editors increasingly allow first personal accounts, particularly in magazine and feature sections. They are usually written by a reporter or ghost written for a person who has undergone a unique experience such as crossing the country alone by hot air balloon. Sometimes a reporter will arrange a unique experience to write about, such as having himself committed to the county jail for a weekend.

Service feature: This is "how to" story, describing anything from how to fix a meal to how to build the cabin. Such features are becoming increasingly popular as news papers strive to get closer to the needs and interests of their readers.

Interview : Although most features are based on interviews, the special interview feature portrays a dialogue between a reporter and another person, often a public figure or celebrity. Sometimes it runs in question and answer format.

String of pearls: This is a "collective" feature, such as a series of anecdotes on a common topic. "Person on the street" interview snippets fall into this category, as would a Valentine's Day feature depicting "the ten greatest love letters of all time. Narrative. Some observers see the narrative as another category of feature, and in its pure form it probably is.

However, the narrative is also a writing technique that can be used in any of the above categories of features. In such forms as anecdotes and case histories, narratives can be inserted into traditionally structured articles.

OTHER TYPES

- 'Soft' news-stories about people on news pages, usually with photographs but not highly topical. These are human interest stories and sometimes referred to in the industry as 'do-ups'.
- First hand experience - used to cover an event or a series of events, and describes the reporter's experience. Investigative journalism falls into this category.
- Straight feature - about an activity, place, group of people, etc.

Writing and Organizing Feature Stories

Feature writers seldom use the inverted-pyramid form. Instead they may write a chronology that builds to a climax at the end, a narrative, a first person article about one of their own experiences or a combination of these. Their stories are held together by a thread, and they often end where the lead started, with a single person or event. Here are the steps typically followed in organizing a feature story:

Choose the theme: The theme is similar to the thesis scholarly paper and provides unity and coherence to the piece. It should not be too broad or too narrow.

Several factors come into play when choosing a theme; Has the story published to the audience before? Is the story of interest to the audience? Does the story have holding power (emotional appeal)? What makes the story worthy of being reported? The theme answers the question, "So what?"

Write a lead that invites an audience into the story

A summary may not be the best lead for a feature. A lead block of one or two paragraphs often begins a feature. Rather than put the news elements of the story in the lead, the feature writer uses the first two or three paragraphs to set a mood to arouse readers, to invite them inside. Then the news peg or the significance of the story is provided in the third or fourth paragraph, the nut graph. Because it explains the reason the story is being written, the nut graph also called the "so what" graph is a vital paragraph in every feature. The nut graph should be high in the story. Do not make readers wait until the 10th or 11th paragraph before telling them what the story is about.

The body provides vital information while it educates, entertains, and emotionally ties an audience to the subject. The ending will wrap up the story and come back to the lead, often with a quotation or a surprising climax. Important components of the body of a feature story are background information, the thread of the story, transition, dialogue, and voice.

Provide vital background information

If appropriate, a paragraph or two of background should be placed high in the story to bring the audience up to date.

Write clear, concise sentences

Sprinkle direct quotations, observations and additional background throughout the story. Paragraphs can be written chronologically or in order of importance.

Use a thread

Connect the beginning, body and conclusion of the story. Because a feature generally runs longer than a news story, it is effective to weave a thread throughout the story, which connects the lead to the body and to the conclusion. This thread can be a single person, an event or a thing it usually highlights the theme.

Use transition

Connect paragraphs With transitional words, paraphrases, and direct quotations. Transition is particularly important in a long feature examining several people or events because it is the tool writers use to move subtly from one person or topic to the next. Transition keeps readers from being jarred by the writing.

Use dialogue when possible

Feature writers, like fiction writers, often use dialogue to keep a story moving. Of course, feature writers cannot make up dialogue; they listen for it during the reporting process. Good dialogue is like good observation in a story, it gives readers strong mental images and keeps them attached to the writing and to the story's key players.

Establish a voice

Another key element that holds a feature together is voice the "signature" or personal style of each writer. Voice is the personality of the writer and can be used to inject colour, tone, and subtle emotional commentary into the story. Voice should be used subtly (unless you're able to make a fetish of it like Hunter S. Thompson). The blatant intrusion of a distinctive voice into news writing has been called gonzo journalism an irresponsible, if entertaining, trend in contemporary writing according to traditionalists.

A feature can trail off like a news story or it can be concluded with a climax. Often, a feature ends where the lead started, with a single person or event.

Some features are written in narrative form. That is, some feature stories begin with an introduction which leads chronologically to a dramatic conclusion. However, most features follow the same structure as the news story.

News stories: News writing aims at providing up-to-date information quickly, clearly and objectively. A particular writing style is used in preparing a news report. The reporter must answer six basic questions: *who, what, when, where, why* and *how* (the five W's and H of news gathering). Journalist and author Rudyard Kipling referred to them as the 'six honest serving men'. Most writers of fiction tell their stories in chronological order developing the characters and story line with detail and colour, but holding the climax until the end. This is called the 'pyramid' style of writing.

The journalist reverses this process and begins with the most important facts in the first or lead paragraph which 'leads' the reader into the rest of the story. The lesser facts and background follow.

There are two reasons for this structure. The first is that when the story is being sub edited for insertion a sub editor can fit the report to a smaller space by cutting from the end without destroying major details. The other reason is that readers can skim the first few paragraphs and feel that information about the most important facts has been read. This style is called the 'inverted pyramid' structure.

A feature writer commonly composes a "soft" news story. This usually contains little hard news so the reporter must write to interest and entertain the reader. Angles are important. The reporter must search for something on which to hang the story. This type of feature story enables the reporter to explore the phraseology of the language. Neat phrases that say a lot in a few words are useful in writing feature stories.

DIFFERENCE BETWEEN NEWS STORIES AND FEATURE STORIES

Follows inverted pyramid form of structure, timeliness is important factor to consider brief note about news, need not general to specific, follows pyramid structure developed not in hurry comprehensive note about news, need good amount of imagination and specific to general.

Hard News and Soft News

A news story can be hard, chronicling as concisely as possible the who, what, where, when, why and how of an event. Or it can be soft, standing back to examine the people, places and things that shape the world, nation or community. Hard news events such as the death of a famous public figure or the plans of city council to raise taxes affect many people, and the primary job of the media is to report them as they happen. Soft news, such as the widespread popularity of tattooing among athletes or the resurgence of interest in perennial gardening, is also reported by the media. Feature stories are often written on these soft news events.

NEWS, TYPES OF NEWS INTRODUCTION

News is both a product and a point of view. As a product, news is gathered, processed, packaged and sold by newspapers, news services, news magazines and other periodicals as well as by radio, television and cubic networks. Today news is also distributed by computers through computer networks like 'CompuServe'. News, in addition is a saleable product. News is what people need or want to know, whatever interests them, whatever adds to their knowledge and understanding of the world around them. News is also what a reporter or an editor at any particular moment considers interesting or exciting or important - something that the reporter or editor thinks will interest or be useful to readers.

News - Meaning

News is the recounting of factual information about events, situation, and ideas including opinions and interpretations-calculated to help people individually to cope with themselves and their environment (Metzler, 1979).

Joseph Pulizer, 19th century publisher of the St. Louis Post - Dispatch and the New York World, had his own definition of news. He instructed his editors and reporters to look for stones were "original, distinctive, romantic, thrilling, unique, curious, quaint, humorous, odd and apt to be talked about".

Turner Catledge, a former Managing Editor of The New York Times, called news "anything you can find out today that you did not know before".

John Chancellor, former senior correspondent for BBC News, suggested that news is a chronicle of conflict and change.

Many editors today are more likely to define news in terms of what their readers want to know - in other words, give the readers what they ask for, whether it is local news of features. Newspaper editors are depending more and more on this definition of news to make their newspapers useful to readers.

Some editors define news as information readers ought to have and need to know in order to get along in this complicated and confusing world.

John B. Bogart, city editor of the New York Sun, who more than a century ago uttered the famous dictum: "When a dog bites a man, that is not news. But when a man bites a dog, that is news". Bogart was dramatically stating a simple principle: News is often the rare and unusual.

Charles Dana, editor of the New York Sun, expressed the idea that "news is what will make people talk".

Stanley Walker, long time city editor of the New York Herald - Tribune, suggested that news is more unpredictable than the winds".

Types of news

Editors sometimes speak of hard news and soft news. Hard news is the serious news: tires, accidents, murders, disasters, deaths, the results of elections, the results of battles, the outcome of wars or the famine. Hard news includes news of severe weather and natural calamities. Soft news is the light stuff; features, human interest stories, news of engagements and weddings, sports, leisure activities and entertainment.

Hard News is generally accepted to mean news of important public events such as actions of government, social condition economics, politics, science, environment, international relations, education, its important, sober and sometimes dull.

Soft News, by contrast, is more interesting but less important personalities, humor, offbeat incidents, individual (and often sensational) crime cases and similar events more noted for reader interest than for significance.

Straight News refers to the presentation of unvarnished facts without attempts to interpret, analyse, or add human interest or "feature" material.

Feature News generally means a story whose primary intent is to entertain or present sidelights rather than take the straight forward informational approach.

Spot News means a sudden happening, such as an accident, a fire, a vote, a court decision, a council action, in contrast to a news account of something that it a long time in the making (such as continuing public debate on, say, whether to build a new city hall). Spot news often means the unexpected.

Chapter – 6

Photojournalism and Reporting

PHOTO JOURNALISM

Few people today could visualize a newspaper without photographs or other illustrations on virtually every page. Photographs increasingly important in recent years help meet the visual competition of television and magazines. The average reader can logically understand a news situation better of pictures help develop it. As the Chinese proverb puts it, "a single picture is often worth for ten thousand words".

The photo journalism encompasses the following:

Taking photo + Taking photo print + Reporting by post or photography.

Photography

Photography is an important tool in the hands of the journalist. It is perhaps this tool which is facing the best and worst effects of modern technology. Advancements in technology during the past four decades have made several cherished skills of the photographer irrelevant. Camera electronics has grown beyond helping him with exposure and focusing. It can now almost fully take over these functions and its dependability is very high. The camera would set for the photographer the best aperture and shutter speed combinations for almost any given situation with or without flash. It can also read the film speed automatically and do the settings accordingly. It is only a matter of years before digital cameras are widely used by the photojournalists. It will free the photojournalist from dark room processing and time consuming manipulations with the enlarger. Wet processing is bound to become a thing of the past, if digital cameras achieve the resolution of films. This may happen any time now. Pictures would reach the newsroom from the photographer in a matter of seconds through telephone lines or wireless. The picture editor can do a quick adjustment of brightness and contrast besides cropping on the computer. And the picture would reach the page in a matter of minutes. If the editor has time, he can use an array of tools available on computer programs like PhotoShop, to improve the picture in many ways.

A picture is worth a thousand words

This old saying will continue to be true. However, the credibility of the thousand words would now be held in doubt from the day you go in for computer manipulation of images. (It is inevitable that you will). Some years back, I reported on a case of the Forest Department employing children for work in their nurseries near Kallar in Thiruvananthapuram district in violation of several laws. I had seen the children working, and the report was carried by THE HINDU which enjoyed a high degree of credibility. Despite that, the forest Minister denied it on the floor of the assembly. He said that the children were there with their mothers who were actually the ones working there (The payments were being made in the name of their mothers). Next day, we carried a follow up showing an overview of the nursery with the caption that no mothers were around. The children were seen working with shovels. Only two grown ups were amidst them. That was evidence which cannot be easily refuted. Even a newspaper with now credibility could use such a picture to prove its point. But the situation has changed now. If mothers are present in such a picture today, they could be easily edited out using latest image processing technology. The background can be changed and the number of children increased, if needed. The credibility of a picture will now dependent on the credibility of the newspaper and the photojournalist. In other words, the credibility of a photograph would be the same as that of the printed word. Thus, the profession losses the strength of one of its tools as capabilities of the tool are enlarged by modern technology.

Besides, the wide scope for manipulation of pictures now throws up several ethical issues. This would range from issues like whether an electric line in the foreground or a disturbing background could be edited out to whether one could change the appearance of a person. Many would countenance the editing out of a disturbing background or the like the what about editing out distances or the like. Some time back, a newspaper published a picture of a policemen running after a few demonstrators with a lathi. The distance between the policeman and the Victims' was drastically cut through computer manipulation. This gave a better effect. Instead of a picture of a policeman chasing away some demonstrators, it looked as if the policeman was about to hit one of them. Is this ethical? In the past also, use of telephoto lenses tended to shrink distances. This was something known to many and they could arrive at a correct assessment of the distances. However, now the situation has reached an unacceptable extreme. Unlike in the case of telephoto lenses, the reader may not be able to guess that the picture has been electronically manipulated. This is true of video footage also.

Similarly, photo finishing, including touching of the negatives, was used by photographers for many decades to improve portraits. Now, there is no limit to how far one can improve or damage a portrait. One can even strip a person of his or her cloths. Where would one set the limits?

Picture editing

Selection of photographs for a weekly or daily newspaper must be done by a person who can evaluate pictures with the eyes of both an artist and technician and who understands news values and page layout. The person who chooses pictures, whatever his little may be realizes that the photographers crave action pictures and editors play up such materials. Many assignments unfortunately, are continued the photographers must use their skill and ingenuity to produce pictures to tell the story.

Handling photographs for a newspaper follows a procedure that is challenging and fascinating. Weekly or daily virtually all newspapers have their own dark rooms and equipment because photographs must be processed quickly to meet deadlines.

Making picture assignment

Editors or reporters can originate photo assignments, but all usually clear through the picture editor, so there will be coordination and no duplication. Most newspapers, large or small, have printed assignment sheets with blanks to provide the information necessary for the photographer.

The person who makes the assignment advises the photographer, via the form, whom to see, exactly where to go when the picture is scheduled and pertinent information regarding why it is being taken. There is also space to show that made the assignment in the event that more information is needed.

Duties of the photographer

The photographers' duties include taking a picture with good composition of with news values that illustrate the story situation. He also must produce a photographic print of excellent technical quality pictures that are too "gray" or contain too much contrast are marks of the amateur and will not reproduce well. When a reporter and photographer are working together the writer obtains the identification of the subjects, according to their positioning the picture. If a photographer is alone, he must obtain the identification. This includes not only the first and last names and where the individuals are in the prints. But also a brief explanations of who they are and what the story situation concerns. A picture without this vital information normally would be unable since deadline present "backtracking" for data.

Training for photographers

Taking news picture is a specialized field, like writing news stories. The techniques of exposure and developing are similar to those of other forms of photography. It should be noted that prints must be clear and sparking. Since newspaper reproduction is of a lower quality than that of top grade magazines where even shadows have considerable depth. Artistic techniques and composition are considerably different from those required in portrait, industrial and commercial or studio photography.

Reporter-Photographers

An area of interest to Journalism new comers is the use of reporter photographers. Commonly called" combination men". Being versed in both skills can be an asset in obtaining a job. Many smaller newspapers do not have the funds to warrant sending both a reporter and a photographer to cover some story situations of even the large ones the frequently prefer combination of men for events, which are a considerable distance away and would require extra manpower for that is often a story of fringe interest. The combination man can fill the dual need covered an area report orderly and photographically. Combination men typically are staff members whose basic skill is reporting and who learn to handle the camera.

Handling pictures

Weekly or daily newspaper can produce many picture situations each usually offering different scenes from which to select. Editors make their selections based on how well a photo tells the story for the availability of space.

Selecting the scene

Picture editors typically would have several methods of making the preliminary scene selection. The oldest way is for the photographer to provide finished prints from the negatives he believes are best. The disadvantages to this method include the fact that the photographer must spend time and materials making prints that will not be used and the possibility that the editor will miss seeing specific scenes which would be more desirable in overall planning.

A highly popular alternate is for the photographer to provide contact or "same size" prints of all negatives. The editor marks the one he wishes to consider and carefully printed enlargement are provided.

Many picture editor can "Read" negatives by holding them to the light and thus visualizing how important points would look in prints thereby making a selection and by passing the time for making contact prints. Cooling at negatives and determining minute points such as facial expressions or other details is an art that one develops through practice.

Negative "readers" are a modern and efficient way of determining what a negative holds. Relatively inexpensive and therefore used at larger dailies, these machines project a positive image of the negative on to a viewer to assist editors in deciding on prints.

Handling the print

Photographers as a rule of thumb provide 8 by 10 inch glory prints. There are standard for handling by editors. Copy runners, artists and printing craftsman since most photographs are "blown" or enlarged from their print size to reproduction dimensions, smaller picture would be likely to show gain after being increased substantially.

Exceptions to the 8 by 10 inch standard include pictorial or other special scenes to be reproduced in extremely large proportions, such photos are often printed on 11 by 14 inch paper to give mechanical technicians a better opportunity for good reproduction. 5 by 7 inch photos will not be blown up when reproduced.

Signing the print

A good photographer would not provide a print dropped so closely that the picture editor would not have room to delete the more of the picture.

The Photo wheel

To determine the dimensions for reproduction, the editor uses a proportion scale that is usually circular in shape and therefore is called the "photowheel' it operated on the same principle as a side rule. Although comparatively new people on the staff of most newspapers can operate the scale. The scale shows by what proportions a photograph will be reduced or enlarged into a ha If tone. Basically it works like this the inside circle (usually coloured) contains numbers to indicate the actual size of the photo and the outer side (usually white) has calibrations to show the reproduction size. All measurements group or down proportionately and the gauge gives exact dimensions in seconds. Inside the inner circle is a gauge which shows the percentage of increase or decrease or from the size of the original. For example a 50 per cent reduction means that the photography will be only one half its actual sizes when reproduced. For a large photo, such a reduction probably would present no problems of the photography or part of a photography being reduced was small in the beginning, a 50 per cent reduction could mean that it would be as small in the newspaper as to be meaningless.

A 200 per cent blow-up would mean that the photography would be twice as big in the newspapers. This increase in size mate little difference for a sharp photography, but if the print were fuzzy or out of focus. These imperfections would be magnified in the reproduction.

Experience teachers the picture editor what to expect in the way of latitude or blow ups or reductions or photos of different quality.

Photographs and picture editing

Write for a magazine or newspaper you need to know something about art. Your sale or placement of a piece of writing may seat on its illustrative potential and may be even on your ability to use a camera. To sell to the small specialized magazines, especially the trade journals you often have to supply photographs "they lend credibility to the subject".

Writer as photographer

You may surprise your self with your success at picture taking. The art involves some of the same judgments needed in writing. Cliftone, Edom, Professors Emeritus University of Missouri, observes that a person who learns to write achieves a discipline whish carries over into his photography. He learns how to select important elements and how to present them. If you don't take pictures for publication you can still use a camera as you use a tape recorder to take notes. The notes would remind you of the appearances of people buildings of scenes.

Photographing in colour

When you move to colour, your job becomes more difficult because you have one more dimension to consider. As much as shape, placement and texture, colour affect your composition, you may want to adjust what your picture includes so that it concentrates on one or two colours instead of offering a rainbow of colours. Some of the best full colour pictures reproduced in magazines appears to be in monotones. At any rate one colour should predominant. Reproducing colour photographs add considerably to a publication's printing costs. Instead of one printing impression, the publication needs four, A few newspapers, printed by the letter press process, get by with three impressions for their full colour photographs, but the photographs look was led out and weak.

To avoid the costs of full colour photography, smaller magazines sometimes turn to duotpnes; photo graphs printed in two colours, one of the colours being black. To do this, the magazine takes an ordinary block and white photography and makes two plates; one or the black ink and one for the colour ink. This is matter divided on by the editor. It does not affect the writer/ photographer who submits black and white for publication.

Processing

Your big problem in photography may be processing you do it yourself you can't do your own full processing you should at least develop the negative your self then turn it over to a processor for prints. If you go outside for processing you will probably try several firms for producing better prints; using a magnifying glass you can pick out the prints that show the most pronounced then ask for blow ups. Editors like to see full frame pints on glory, smooth paper. You still leave while space on the outside so that the editor can make his crop marks.

Caption writing

Always ask yourself who? Why? When? And how? And make sure that none of them remains unanswered in your caption.

Retouching packing of copying

Generally the photographer themselves retouch the photographs. There are several methods at retouching-pencil, scrape-blade, various dyes and water colour paints, rejoining or print surgery was made with scissors and paint.

REPORTING - QUALITIES OF A GOOD REPORTER, REPORTING SKILLS

Reporting Vs Reporter

A reporter Versus one who observes the passing show in the widest sense of the word and pictures and its details for the benefit of the whole society. A reporter may be accurate, conscientious, a good citizen and take part in moulding the views of their eppie, but he cannot play his part successfully unless he keeps his eyes open and his mind attuned to the present, future as well as the past. He is a leader of men in many senses of the word. He must not express his own views in what he writes-he must know most sets of circumstances-but everything he writes must express his mind and its condition. He holds up a mirror and how much clouded or clear it is, depends on the truth or a twist of the truth which he makes in accordance with his nature and mental equipment.

The Role of a Reporter in a Newspaper

A good reporter seldom sticks to a newspaper for long. Usually he passes on to news agencies or get promotion to look after other aspects of newspaper production like news editor or a chief correspondent. A reporter's mind is like a sponge, paying a good deal of attention to purely mundane things but learning something every day and cleaning his mind of matters not up to the mark. He must organize his knowledge and conduct. He must understand the principles of government in general and in some details. He must know general history, particularly the history of his area. Though he can pick up these things as he goes along his duty, it is better for him to supplement his knowledge by a planned study.

Duties of a Reporter

A reporter is a gatherer of news, and as such performs an important function in a newspaper establishment. He has to interview persons and attend public functions and meetings, press conferences and law courts to investigate events of public interest, to collect news and to ascertain news on contemporary events.

His work changes daily; as such he should be prepared to handle any assignment and move anywhere. He should have special knack of meeting all sorts of people in all type of circumstances. He must possess abundant self confidence, so as not to be overawed by the rank or position of an individual. He should be able to grasp the situation quickly and reduce it into writing in the shortest time and in a readable form. The narrative should be attractive so that the readers should enjoy reading it. He must have an eye for the important discussion/decision which are to be included in the report, omitting all other unimportant / irrelevant matter He must be temperamentally so framed that he does not get irritated, even if at times he has to wait for hours to meet an important person or come across an event. When a newsworthy occasion does come he should be able to grasp it quickly, and write it out with great speed.

Major reporting Departments

The major reporting departments of a newspaper are:

Local Administration and District courts

The International Field

The Art Department

Sports

Political Affairs including Parliament

Economic Affairs together with Trade and Commerce

Book Reviews

Science and Technology

Industry

Qualities of a good Reporter Memory for faces

A young reporter must have good memory for faces. He must cultivate the sixth sense and grow particular brain ceils assiduously if he is to make Hill use of his opportunities. Once he is out to achieve this he will find plenty of people in his circle who are valuable to him as news supply source.

Remembering from pictures

A successful reporter must have good memory Jar-pictured faces. He must also take out information about the personality he wishes to approach for interviews or news.

A good memory for pictures is of tremendous utility to a journalist in a strange environment. Every journalist has the experience of going from pillar to post for information or confirmation when minutes and seconds keep flying past He may be sometimes held up by the official folk unnecessarily. What he should do is to care for every precious moment if he wishes to be the first to get the news through. But he has got to do everything so diplomatically that he does not lose any friend who can be of use as a source of news.

Making friendships

It is important for the reporting journalist to make right type of friendships with his sources of information or news. He will find his source of knowledge widened or restricted, according to the way in which he can cultivate the friendships. Sometimes, it is alleged that journalists are too busy to keep a social life. In actual fact, the case is the reverse. A journalist touches more of the social circles than any man. He remains on the fringe of them all and can turn them to his own professional use if he has the necessary capability.

Freshness of mind

A journalist cannot afford IQ be stale in his ideas, concepts or pursuits. He has to be on his mental toes all the time. He must do a lot of reading and writing be honest to himself. Find out more quickly than others the precious little knowledge he is in search. His writings must emit the bright sparks with every new contact. His every new story must be an improvement over the earlier one.

Thrill of the hunt for news

An ideal reporter must respond to the urge and thrill of the news. News getting is not a matter of routine. He must be dynamic in his approach so that he can get preference to the routine journalists from the news agencies or other newspaper that my compete with him to find access; to the source of news. Unless he has a thrill for news, he cannot prove to be a more efficient news gatherer and interpreter of the movements of which he is an observer and from which he has to build the background for future coverage. The competitive aspect of modern journalism adds to the thrill rather than subtracts. The just for a new fact in competition with others must have the effect of throwing a heightened light upon it and of getting it out of focus in relation to its setting. Thus, there is a constant war between the journalist's conception of the public appetite and the need to weigh facts carefully and to keep a sense of values which is intrinsic.

Sense of Responsibility

An ideal reporter, however brilliant and imaginative, must maintain a sense of responsibility. He must adjust his balance of things in such a way that he can keep tothe interests of his organization without sacrificing the public good. His over enthusiasm should not land him into a situation where he may be called irresponsible. He must realize that he holds in his hand one of the reins by which mass movements are forged and that he must use it purposefully and with care.

EDITING - GENERAL PRINCIPLES OF EDITING

The task of making changes arid corrections in the body of the story before it is set in type (Crump, 1974). Ahuja (1975) defines it as a process by which the editorial staff provides material for publication and prepares the subject matter called "copy" for the printer and the engraver. Editing is nothing but deciding what news shall appear where and how, apportion space, assign priorities and decide now certain stories should be featured or displayed (Tampi, 1990).

General principles of editorial

Editorial or review column should still be objective in the sense that its conclusions are based on fact and reflect the writer's argument rather than his personality

First person singular should be avoided as it can weaken the criticism by making the critic seen unsure of himself

Avoid lack of logic

Should be complete

Avoid vagueness

General principles of editing

a. Dynamic role

1. The functions and responsibilities of the editor keep on changing
2. The editor in order to maintain and increase the circulation day to day, has to compute the public mind and express opinion in the light of popular trends prevailing from time to time.

b. Public relation duties

The editor of a newspaper must have patience, human sympathy, a working acquaintance with type and how to use it.

c. Field duties

He must keep in touch with public opinion by mixing with the readers in every walk of life.

d. Round the clock routine'

From early morning to midnight, the editor is on duty either in the office or on his round to visits and even then he is certain that his sleep will not be disturbed with the arrival of a vital piece of news.

e. Writing of leading articles

The editor himself should write leading articles (or) whether he himself should confine himself to briefing other members of his staff.

f. He must have an infinite amount of patience and a keen interest in news of all kind.

g. He must have a good general educational background with a fair amount of historical, political and economic knowledge.

h. He must be a good mixer, he must be on the lookout for news all the time, he must learn to scan the newspaper.

II. According to the Author Mehta (1969)

1. Paying attention to language

The editor is expected to pay much of his attention in understanding language and applying it.

a. The study must be precise and to the point

b. It should be clear enough giving no room for ambiguity

c. Pace should be appropriate to the content. Pace can be determined by the sentence length. Sensational and breaking news should not have sentences of long lengths.

It may irritate the reader. The story should have transitional devices that lead the reader from one thought to another. The study should appeal to the reader's senses. He should keep changing to keep pace with the modern world.

2. Rewriting

Rewriting enhances the quality of output.

3. Writing headlines

The rules followed to write headlines are

1. It must carry an active verb
2. Should be in the present tense
3. Avoid abbreviations as far as possible
4. Comma can be used in place of and

Editing - General Principles of Editing

The task of making changes and corrections in the body of the story before it is set in type (Crump, 1974), Ahuja (1975) defines it as a process by which the editorial staff provides material for publication prepares the subject matter called "copy" for the printer and the engraver.

Editing is nothing but deciding what news shall appear where and how, apportion space, assign priorities and decide now certain stories should be featured or displayed (Tampi, 1990).

General principles of editorial

Editorial or review column should still be objective in the sense that its conclusions are based on fact and reflect the writer's argument rather than his personality.

First person singular should be avoided as it can weaken the criticism by making the critic seen unsure of himself

Avoid lack of logic

Should be complete

Avoid vagueness

General principles of editing - Dynamic role

1. The functions and responsibilities of the editor keep on changing

2. The editor in order to maintain and increase the circulation day to day, has to compute the public mind and express opinion in the light of popular trends prevailing from time to rime

b. Public relation, duties

The editor of a newspaper must have patience, human sympathy, a workingacquaintance with type and how to use it.

c. Field duties

He must keep in touch with public opinion by fixing with the readers in every walk of life

4. Proof reading

Proof reading is an important part of editing writers who took interest and took part earnestly in the proofreading process.

5. Layout and design

Is the matter plan based on which the make-up of a page is carried out.

Basic of Editing

1. Abstraction!

Avoid lengthy abstractions particularly those using 'basis' and 'situation': on a regular basis' ... on an individual basis' ... on a weekly basis' quite simply with 'regularly' 'individually' every week'.

2. Acronyms and Abbreviations

""The first mention of an organisation, examination, curricular area and so forth must be given in full, followed by the abbreviation in brackets.

No explanation is required for household terms such is BBC, EU, UNKSCO or very well known educational acronyms (GCSE etc.)

3. Adverbs

Adverbs of time such as frequently, often, sometimes, rarely, never are in some cases and inappropriately, giving a false impression of inspection over a quite lengthy period instead of only three or four days :

The teaching in physics lessons was rarely less than satisfactory'. Better 'In only a few physics lessons was the teaching less than satisfactory.

4. Capitals

Naming of individuals, Groups and Curriculum subjects in capitals.

Use capitals for proper names and specific titles, but not for general school appointments: St. Nemo's College the Founder's Day ceremonies the headmistress

5. Guidance on Upper and Lower case

Use lower case when referring to: individuals as defined by their roles; subjects; parts of the school. E.g. Headmaster/ mistress chemistry laboratory

But in the following cases use CAPITALS for clarity when the terms are used by the school as proper names:

E.g. "Girls proceed from Lower School to Upper School..."

'Children in Nursery and Reception ...'

6. Choice of Words

The style in which a report is written has to be clear in meaning and, on the whole, plain. Unduly lengthy, cumbersome sentences should be avoided; on the other hand too many short, staccoto statements often make it difficult for a reader to follow the thread of an argument. Avoid words that readers may interpret either as praise as criticism.

7. Compression

Sometimes 'compressed thinking' may muddle the message.

Eg: The continuing pressure on the curriculum for all pupils cannot realistically support eight periods for classics'.

8. Conjunctions

While, whilst whereas and although have the force of despite the face that and, like but, indicate a contrast. They are best placed at the beginning of a sentence. Whereas though and although are interchangeable in this position, though may sometimes be used in mid sentence.

9. Hyphenation

Hyphens are in most cases used to join two words in a single related idea where the two words, used separately, could mean something rather different. One of their main purposes, therefore, is to avoid ambiguity.

a. Hyphen of Readability

Certain prefixes such as anti-, co-, micro-, pre-, pro-, semi- etc. are usually hyphenated.

b. Hyphen in Word-plus-word combinations Compound Nouns

Hyphens should generally be avoided in favor of one word or two. A few examples are: staff room, boardroom.

c. Number between 21 and 99 are always hyphenated; twenty three pupils in Year 8. Fractions are shown as one third of the teaching staff. Three quarters of the total curricular time.

d. Fix I phrases such as one-to-one; give and take; full time; part time take hyphens.

e. Adverb + Past participate The adverbs well, better, best, ill, worse, worst, little, often, seldom, much followed by a participate are always hyphenated when they precede a noun but not when they follow the verb, as in the following examples :

A well planned lesson The best presented work

f. Some combinations are hyphenated in all positions, before and after the verb

Adjective + adjective red –hot Adjective + participate harsh-sounding

10. 'Nostrums', self-evident truths and preaching

While all judgement presupposes a certain view of what constitutes good or poor practice, it is inadvisable to preach. Better - and less pompous - to write 'Many classrooms had not been properly cleaned' than 'It is important that classroom should be properly cleaned.

11. Nouns

Beware the frequent use of nouns as adjectives. There is usually a perfectly adequate adjective or possessive. Avoid 'curriculum development' and 'pupil development', 'curricular development' and 'curricular development' and 'pupils' 'development' arc preferable.

12. 'Number'

When preceded by 'the', the word number takes a singular verb: 'The number of grade A results in high' When preceded by 'a[1] takes a plural verb: 'A number of solutions have been suggested'

13. Numbers

In general express numbers upto nine in words. 10 and above in figures. But do not mix in close proximity. Of the nineteen classrooms, there are too small and twelve are isolated'. At the beginning of a sentences, write numbers in full: 'Fifty-eight pupils are boarders ....'

Percentage and ratios need not usually be expressed more than one decimal place:

'Five per cent' or '8.8%' when referring to proportions (of satisfactory/unsatisfactory lessons; of boys and girls in the school; of the teaching experience of staff), use simple fractions in order to avoid giving an erroneous impression of precision : 'Just over two-thirds of the lessons were at least satisfactory . . .' (not 69%)

14. Objectivity

An ISI report should present in ***impartial and objective*** account of observations, judgement and recommendations. The forms of expression used must reflect this objectivity and eschew any phrasing which implies partiality, personal opinion or bias on the part of the inspection team.

The strict rule is that ***remarks in the first person (singular or plural) must NOT*** *be made.* Forms which are debarred include not only; 'I did not see any sign of marking were impressed by . . .', but also 'in out- view'.

15. Punctuation

Commas are a matter of some contention, but writers will recognize that in this present sentence the clause in parenthesis must be defined by a pair of commas.

Semicolons need to be used sparingly. They may join two or more main clauses which, logically, are closely associated and whose relationship might not be so obvious if separated by full stops.

Colons are used to introduce exemplifications, expansions or illustrations of what precedes the colon.

16. Singular and Plural

'Virtual plurals' are singular nouns that nevertheless describe a multiplicity. 'Staff is the most common of these items.

'Virtual singular Some words though plural in deviation, have come to be regarded as singular, 'data, in the sense of a collection of quantitative information, can take a singular verb.

17. Tenses

The golden rules for the use of past and present lenses are:

Use the present tense for long-term and permanent features

Buildings, organization and staffing, systems of assessment and pastoral care, resources, curriculum documents; these form the general setting.

Use the present-tense in general evaluative comments of standards, teaching, pupil's progress, learning and behaviour which refer to the entirety of what was seen during the inspection.

In full subject reports, use the present tense to describe attainment, progress, learning and teaching a well as resources etc.

Use either the present or a past tense to describe attainment in tests / examinations, depending on the precise phraseology. GCSE results are above the national average for all schools'

Use the past tense for assessing the quality of teaching and learning in relation to specific lessons or other activities were observed. In a year 3 lesson the pupils lacked concentration'.

18. There is / are

Avoid excessive use of there is .... there are .... there is a .... which .. . .'

While these phrases are occasionally unavoidable and sometimes essential, repetitive use — it tiresome. In most cases the meaning can be expressed more directly, for example:

There is a lack of suitable textbooks in history which needs to be addressed'. **Plain** Language Editing Strategies

Strategy	Description
Reduce excessive length	Reduce wordiness and remove irrelevant material
Use common words	Eliminate unusual or low frequency words and replace with common words (e.g., replace "utilize" with "use")
Avoid ambiguous words	For example, "crane" should be avoided because it could be a bird or a piece of heavy machinery
Avoid irregularly spelled words	Examples of irregularly spelled words are "trough" and "feign"
Avoid proper names	Replace proper names with simple common names such as first names

Avoid inconsistent naming arid graphic conventions	Avoid multiple names for the same concept. Be consistent in the use of typeface
Avoid unclear signals about now to direct attention	Well designed heading and graphic arrangement can convey information about the relative importance of information and order in which it should be considered

NEWS PAPER BALANCE

Nature may be said to abhor lopsidedness almost as much as it does a vacuum. Branches grow on all sides of trees, animals have two eyes and normal human beings have feet of comparable weight and size. It is literally natural, therefore, that a basic principle of art should be balance and symmetry, which is the balance of allparts with reference to the center of the whole.

Use of a definite plan for every page of a news paper, no matter how small, has an unquestioned value. The well-dressed newspaper always commands respect; the slovenly one forfeits it.

If colour i.e., the degree of blackness of typographic elements- be thought of as weight, the two kinds of balanced makeup become understandable. One balanced on an optical fulcrum in the center of the page, establishing equipoise between headlines and cuts on either side. The other has an off-center balance, with an optical fulcrum near the heavier unit. And thus we have named the two general systems of balanced makeup

1. Center balance
2. Off-center balance

1. Center-Balance Makeup

There are two common patterns. One is often called balance makeup, but, to avoid confusion in terms might well be designated true balance. The other is inverted pyramid makeup. Both strive for symmetry and typographic beauty through an orderly balanced use of headlines, boxed heads and cuts.

In true-balance makeup: a large headline placed in the last column of the front page calls for a correspondingly large one in the first column. Other headlines are arranged in an orderly sequence to carry out the impression of equipoise. If there are headlines below the fold of the paper, they also are balanced. If there is a one column cut at the top of column 2 it is balanced by a cut of like size at the top of the next to the last column.

Conservative newspapers, especially, favor true balance makeup, or some variant of it. However, there are but few that preserve a strict hair's breadth balance, as this is likely to become monotonous to readers. The most familiar variation is to place a three column cartoon at the tops of columns 3, 4 and 5, or 6, and run major headlines in columns 1, 6 and 8, or 1, 3, and 8 with subsidiary headlines in the remaining columns and beneath the cartoon. Carefully selected fillers take up the remaining space. Another variation is the use of unusually large headlines in columns 1 and 8, or in column 8 only. In other cases there may be two column headlines toward the bottom of columns 2 and 3 and 6 and 7.

II. Pyramid makeup is similar practices are followed, but with one difference. An optical fulcrum is established in the page's center near the fold, and heads? cuts are so arranged that they create lines of an inverted pyramid. Using a crayon to connect color elements on a page will help the student in locating the optical fulcrum and classifying the makeup system employed.

2. Off-Center-Balance Makeup

The two makeup patterns just described have an obvious balance; more subtle is the balanced effect secured by the brace and the broken column methods which use the off center principle.

I. The brace pattern. It developed because of the common need to give to one outstanding story a maximum display. Once streamer lines were accepted, it was easy to give one to the big story, and then drop into a large headline at the top of the right and columns. The typographic weight was thus concentrated in the upper right quarter of the page, roughly suggesting the pattern of a brace or wall bracket, hence the name.

Variants of brace makeup include the occasional reversing of the plan-that is, leading from the streamer into the columns at the left. Brace makeup is especially useful for papers that use three or four column heads instead of streamers.

There is no rule for achieving balance in brace makeup. Various devices, such as two column heads, boxes and cuts, are spotted elsewhere on the page to offset the concentrated weight. It is in solving such problem that new headlines and new ways of displaying news values are evolved and editorial acumen and ingenuity are shown at their best.

Brace makeup of the Washington post gives the page an off-center balance. Two column headlines on the left are used to offset the weight at the right of tne page. The two pictures and the banner headlines give the page its brace effect. The two display stories at the bottom of the page keep the page from becoming top-heavy. One problem in this double banner arrangement is that the reader has *to* hunt for the story accompanying the second banner headline.

II. Broken-column makeup : This calls for typographic skill and a nice sense of symmetry and proportion. Otherwise the page may look like something assembled by a printer's apprentice assisted by the office boy. It can be done expertly, however, giving the news display an effervescent quality.

If a news index is run on the front page, the tendency is to run fairly long stories so as to reduce the necessity for trailing jumps through other pages. But broken-column makeup was developed primarily to get as many short stories on page 1 as possible.

An impression of symmetry on a page out up by many headlines, boxes and cuts is difficult to secure. But it can be done by carefully arranging all elements so that they do not cancel each other out by propinquity and by spotting the large ones where they tend to give a pleasing pattern to the whole page.

3. Unbalanced makeup

Just as flush headlines mark a reaction against the carefully organized headlines that stress symmetry and design, so some newspapers refuse to be cabin'd and confined by the limitations upon freedom imposed by balanced makeup. When fred of arbitrary restraint, makeup takes multitudinous forms, but two general trends are discernible and can be given the labels circus and panel.

Balance will suggest poise and equanimity-which are precisely what certain newspapers do not want. Symmetrical niceties, they theorize, make for uneventful reading; do not jolt the reader along from one sensation to another. So, carrying the broken-column technique to its ultimate, they have produced the circus type of makeup.

Chapter - 7

Over view of Farm Journalism

Journalism (Crump 1974)

Journalism encompasses fields ranging from reporting with words, photographs, editing and from newspapers to television.

Journalism is the key to communication. Until 1920 Journalism was confined to writing for news papers and magazines. Today Journalism include not only news papers and magazines but also radio, TV, cassettes, public relations, advertising, book publishing, education and cyber journalism.

Journalists (Crump 1974) are keeping eyes, ears at the curiosity of the public 'and must be so broad in their outlook that they can translate events in many fields.

Journalism (Ahuja 1987) is the part of social activity concerned with the dissemination of news and views about the society.

As quoted by Dahama (1973) according to welsely and Campbell Journalism is the systematic and reliable dissemination of public information, public opinion and entertainment by modern mass media.

Agricultural Journalism (Ward 1959) is the timely reporting and editing with words and photographs of agricultural news and information for newspapers, magazines, radio, television and net.

News (Crump 1974)

It is a great sense, in the break down of human relationship and the occurrence of events which are expected and even surprising.

Hardnews

Stories of primary or urgent importance rate the hardnews classification. There situation would provide "must" stories that readers would expect to see.

Softnews

News of secondary importance is labelled "soft" news and might concern such items as a school or church, social meeting etc.

Farm news

It is any information which is new, significant, timely and of interest to all those connected with farming.

News is information about an event, some departmental plans, movement of important persons and announcement of Government policies etc.

Report / Reporting

It is giving of information periodically about something seen/ heard/ gathered to appropriate agencies.

Editing

It is the processing or arranging manuscript i.e., (Unprocessed first report) by selecting and sorting for the proper convey of information with accuracy and relevance.

Scope of Journalism: (Dahama – 1973)

The scope of journalism can be discussed in terms of perspective of mass media of communication as follows.

1. Many shall run to and fro and knowledge shall be developed.
2. Men hungry for the knowledge of past.
3. Men hungry for the knowledge for the present.
4. Modern men are the news paper men.
5. Modern men reading periodicals.
6. Spoken word as factor in mass communication

Purpose and functions of Farm journalism

Explorers of Farm Journalism discover that media of mass communication have one to three purpose.

1. They may inform about agricultural development, disseminate news and miscellaneous news on agriculture and allied fields like animal husbandry, horticulture, fisheries, forestry etc.

2. They may influence, giving the farming community either a social or commercial message, which may be applicable for societal development.
3. They may be entertainment, presenting features, fictions, humour, comics and similar matters
4. In addition to this the voice of the farm people and watching of their interest interms of crop cultivation marketing etc.

A. Role of Agricultural journalism

1. To make people aware about the innovation that are available in various fields like agriculture, animal husbandry, fisheries, home science, child care, health care, family welfare etc.
2. Serves as a means to know about the surrounding and happenings and developments around us. (e.g.) demonstration, extension activities, approaches, epidemics etc.
3. Provides sources to be contacted in the case of inputs suppliers including institution that concentrates on specific target group
4. Provides advance information for proper planning in the case of crop cultivation etc.
5. Make the people aware of national policy with regard to the agricultural department programmes and the role of farmers in implementing such programmes.
6. It acts as a instrument in creating social change in rural settings.

Farm journalism provides farm related information to beneficiaries such as consumers, legislators, home makers, farmers, agri business, commodity groups and public sector agencies.

Farm journalism in India is communicated through two ways viz., publication through farm magazines and farm pages of dailies and agricultural publications. The former is published by SAUS, ICAR institutes and established societies of scientist in their disciplines concerned. The major output interms of results of research work, experiences of researchers/ field workers, feed back from their field visits etc., which largely satisfies the scientific community. In order to satisfy the requirements of beneficiaries journals have been published in local language. The content will be scientific information, success stories, question and answers etc.

Journalism in earlier days did not cover the farm and related aspects. The pioneer in this area was scientists from faculty of agricultural extension. The job of the personnel working in agricultural extension is to transfer the technologies, generated from different subjects to the beneficiaries, based on their need. Also, to get feed back about those delivered technologies, though it is one of the three fold functions of SAU scientists, the proportion of time spend by extension scientists is more than compared to others. Thus the channel they are using other than extension methods are through farm journals / magazines and through special papers in dailies on specific days.

Expectations of Farm Journal readers

An innovative client will expect the following information through farm journals.

- All events related to farms
- Information on Package of practices
- New varieties / Technologies
- Integrated nutrient Management
- Integrated weed Management
- Integrated Pest Management
- Post harvest Technologies
- Good Agricultural practices (GAP)
- Value Oriented Agricultural Production (VOAP)
- Marketing channels / Techniques
- Success / experience of farmers stories
- Farm policies
- Farm programmes
- Farm Entertainment
- Trend in global farming

Genesis of Farm Journalism

1910 - Bengalis Monthly '**krishak**' (the farmer)

1910 - Krishi sampad published from Dacca (Now Bangladesh)

1924	-	**'Sonar Bangla'** from Kolkala
1931	-	**'Kirishi Dakshini'**, Kolkata
1931	-	Agriculture and Livestock (English) Imperial council of Agricultural Research
1934	-	**'Krishi katha'** (Agricultural News), Bengal
1935	-	**'Krishi Lakshmi'**
1947	-	Kheti (Hindi) ICAR
1955	-	**'Gram sevaker chitti'** (letters of VLWN) by Agricultural and Community Development Department
1963	-	News letter or tabloid of agricultural News paper
1962	-	Krishi Pragati, Monthly farm journal (Bengali)

Role of Farm Journal as Communicator in Extension

As discussed earlier, communication is an effective process in extension, develops awareness and inquisitiveness among target gropus. Three types of communication, human exists, embodied, media-embodied communication disembodies communication. Among disembodied communications, 'Farm Journal' plays a pivotal role since it reduces gaps between laboratory and land, starting from awareness level to adoption stages in extension.

The process of extension commences with communication, acting as driving force, dealing with mental phenomena. It develops awareness, stimulates inquisitiveness, crystallizes decision and energizes action of the target group, farmers. Berlo (1960) visualized communication as a process while Ban and Hawkings transmitted from source to receiver. Roger and Shoemaker (1971) developed communication model as

S (Source)– M (Message)–C (Channel)-R (Receiver)–E (Effect).

Principles

Extension communication should be guided by,

1. Be sure of what is to be communicated.
2. Be careful in use of language understandable to the target group.
3. Be sure about the right climate.
4. Be sure about the right medium.
5. Be certain about the quantum of communication.

6. Be sure the message communicated properly.
7. Be sure of the voice of the communicator.
8. Be sure that the receiver motivated.
9. Be sure that action should not contradict communication.
10. Remember communication is a two-way process.

Importance of Farm Journal in Extension

Farm journal is a written form of effective communication. It carries message from source to the farmers. Leaflet, bulletin, folder, news letter, etc. are generally published by the research institutions or by the concerned department of an organization. Such publications cover recommendation, technology, package of practices of rural industries. All these do not carry the view of the target group. The farm journals keep space for publishing information relating to rural industries.

Farm journals publish information in sporadic way. Extension workers may find difficulties to develop rapport with the editor of newspaper. As a result there occurs wide gap between the editor and the institutions. The farm has to play great role under such circumstances. The farm journal must cover documentation, communication, motivation, coordination, popularization and advertisement, etc. as given below:

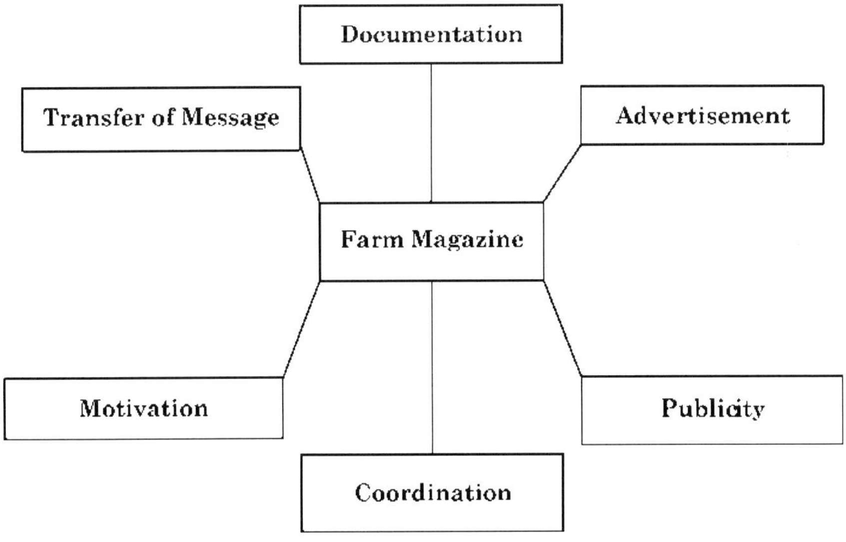

(a) The farm journal is bestowed upon the responsibilities of transferring message from source the receiver. It has to develop documentation system which is very effective when it maintains the list of resource persons and keeps close relation with national and international documentation sentries.

(b) Supe (1987) visualized the problems of researchers to meet the farming community in one hand and the difficulties of the farming people to attend the research institutions to get informations relating solutions of their problems on the other hand as follows for communication:

- ❖ It can publish extension activities, extension recommendations, package of practices, etc.
- ❖ It can highlight achievements, success stories, technological impacts, etc.
- ❖ It can illustrate opinions of farming communities, front line workers, subject matter specialists, etc.
- ❖ It can cover government policies, schemes, institutional supports, etc.

(c) The purpose of farm journal is to develop awareness leading to motivation at two levels – intrinsic motivation and extrinsic motivation. The impact of intrinsic motivation is almost permanent where the farm journal takes the leadership in covering the largest group. In this motivation process, the farm journal may publish suitable package of practices to satisfy the need of the farming community.

(d) The gap between lab to land is due to improper technology transfer. The system of technology transfer requires coordination between workers, front line workers, innovators, financial institutions, government and non-government organizations (NGOs). Developing coordination by the farm journal can intensify the process of technology transfer.

(e) There is a great need of popularization of high yielding varieties of crops, improved materials/inputs and methods of implementation for lab to land programme. The farm journal should publish the success stories, technological impacts, etc. for the purpose.

(f) Advertisement is an important method by which manufacturers and customers are brought close to each other. The objective of advertisement is to reduce the gap between the producers and farmers of the materials/inputs. The Agro-based industries are now developing quick disposal of their products such as agricultural machineries, farm implements, fertilizers and other agro-chemicals. The farming community requires such important inputs to adopt improved technology in farming. The manufacturers and the customers could be successfully related by farm journal by highlighting these products of manufacturers through advertisement in the farm journal.

Factors Affecting Dissemination of Agricultural information in Extension Programme. In any agricultural extension programme, effective dissemination of agricultural information is as important as the development of technical agricultural knowledge. Extension workers must be equipped with technical knowledge and they should be able to pass it to the farmers for whom it is meant. Before 1955, none of the Agricultural Colleges/ Universities was teaching subjects like Agricultural Extension and Rural Sociology. In the absence of such knowledge, our Extension Workers remain ineffective.

Dissemination of Agricultural Information requires

1. Well-trained Extension Workers who are to use their knowledge of improved practices on agriculture, livestock production, poultry, fishery, home management, etc. under village conditions.
2. Motivation in the Extension workers/agents so that they are able to do their job with zeal, enthusiasm and high morale.
3. Approach of the Extension Agent to:
    a. Understand and subscribe to purposes in the Extension programme.
    b. Understand the culture of the farmers, their problems, resources and others for their development.
    c. Help them recognize their problems.
    d. Motivate them to adopt improved practices.
    e. Demonstrate them in a way so that they are able to do the job themselves.

f. Coordinate and co-operate with specialists (SMS) and other agencies required in the Extension programme and use them as resource persons for the work.

g. Recognize the real leaders in the villages and use them for the work programme.

h. Evaluate the work and explain it in the meetings in an understandable way to the farmers, and make adjustments according to the situation, keep the morale of farmers high and encourage them for adopting more and more improved package of practices when they adopt some of them.

Jana, B.L and K.P. Mitra, (2105) Farm Journalism, Agrotech Publishing, Academy, Udaipur, ISBN: 81-85680-98-1.

II CYBER EXTENSION

Chapter – 8

Genesis of Instructional Technologies

GENESIS WITH DEVELOPMENT OF INSTRUCTIONAL TECHNOLOGIES

History of International Technology

International Technology was born as a military response to the problems of food shortage during the World War II in United States. There was a definite need to fill the factories with skilled labour. IT provided a methodology for training in a systematic and efficient manner. With this came the use of highly structured manuals. Instructional films and standardized tests.

IT Today

IT is continually growing field of study and practice utilizing technology as a way to solve educational challenges posed by villages. For example many universities are embracing its IT methods to help increase enrollment while decreasing overhead costs. Other IT tools provide opportunities for students to interact with experts in field, even if they are not located physically close to each other.

However, while IT promises many solutions to educational problems, pushback from teachers and administrators to the use of technology in the classroom is not usual.

However, this could be considered naturally as a subconscious or conscious realization that the ultimate aim of IT is to reduce human element of instruction to its barest necessities. The counter argument for instructional technology enthusiasts is that human interaction will always need to be taught and human teachers will continue to be needed us socialization trainers.

1. Computer Aided Instructional Technology (CAI)

Computers find their applications in many fields say, Engineering, Agriculture, Architecture, Cinema and Bioinformatics etc and also it find its application in instrumental technology. CAI has the tremendous potential for delivering instruction. CIA has potential to revolutionize the teaching and training process. The most exciting innovation in educational technology is CAI.

The use of computer directly in teaching and learning is called Computer Aided Instruction.

Characteristics

- As instructor – led computer based teaching aid
- Cost effective, fully interactive
- Difficult concepts are made easier to understand (Instructor Taught Lesson)
- Enhances retention of knowledge
- Teaching content of courses can be reduced
- CAI transmits difficult concepts in an easy to understand, entertaining way, alleviate language difficulties
- Greater standardization of instruction. There will be correlation equally half to the teaching.

Advantages of CAI

- Learner can learn on his own. Students who experience difficulty can move at a slower speed, review some trouble some sections and request additional practice until the material is mastered.
- Active learning
- Variety of learning subjects
- Record keeping
- Flexibility
- Timeliness
- Visual reinforcement

2. Video Text

Video text refers to any interface electronic system that allows the users to send and receive message on data from either a PC or a dedicated terminal.

Videotext works with the help of the public telephone connection and the domestic TV set or computer screen form the integral components of a videotext system.

Consumers provided with videotext facilities can make use of it for varied applications. Services provided by videotext fall in one of the 3 areas.

1. Information retrieval services such as obtaining stock prices and weather forecast.
2. Transactional message services which enable the purchase of merchandise over the network (e-shopping).
3. Interpersonal message exchanges which may include conferencing, chat channels or electronic mail.

Videotext is often referred to as online service or interactive network. It differs from the broadcast media delivery systems due to the special qualities of interactivity endangered by the technology, which allows the user to personalize his media use rather than act as a passive member of an aggregate audience.

3. Communication Satellites

Communication Satellites are owned by various individual nations as well as collectivelly by group of nations to cater to several applications like broadcasting, television and telecommunication.

INTELSAT (International Telecommunications Satellites Consortium) is an international body consisting of more than 90 member countries of which India is one of the members. The satellite commissioned by INTELSAT caters to the collective needs of member countries as per the mutually agreed tariff structure. Ground terminals (earth stations) for sending and receiving signals from the satellites are located in several places in the member countries. The earth stations for India are located at Pune and Dehradun.

Satellite Communication in India

India is credited with taking several major initiatives in the field of satellite communications. The famous SITE (Satellite Instructional Television Experiment), during 1975-76, in which about 2400 remote villages in the country were severed with television via satellite.

This experiment has proved the efficiency of satellite communication as a cost effective means for countrywide dissemination of educational and development TV programme. Gradually our country has developed its own satellite system, INSAT (Indian National Satellite), INSAT is a multipurpose satellite service catering to telecommunication, radio and TV transmission and forecasting. The second generations of INSAT satellites, which are currently in operation, have a larger number of transports, which can be used for further expansion of satellite communication.

INSAT	Year of launch
INSAT-1A	1982
INSAT-EB	1989
INSAT-D	1990

In general, regional telecasting can be extended in all language zones and new series like teleconferencing is introduced.

Direct Broadcast Satellites (DBS)

DBS are specially designed for broadcast applications so that home receivers equipped with a small dish antenna and an interfacing unit can directly receive the signal transmitted from them.

4. Teletext

The videotext transmission based on broadcast signals is known as teletext. Teletext is a system that links a computer to a television, by which text and graphical information can be transmitted on a one-way basis to home viewrs.

Indian television joined the advanced nations in 1985, when it started the teletext known as the Intex service to telecast latest news and information on the stock exchange, airlines, railway timings, weather information etc.

So teletext is a from of broadcast technology by means of which several pages of textual information can be transmitted on an already existing television channel.

5. Interactive Videotext

The videotext, which is transmitted via telephone cables, data line or private network, is called interactive videotext. It is a fully interactive system and can print a hard copy of any page that appears.

6. Microcomputers

It was introduced in 1978.

- It can process all kinds of information and keep the record to be retrieved as a when needed.
- It can publish information material at reasonable cost for extension purpuse.
- Microcomputers can communication with each other if linked together with the help of certain hardware like microcomputer modem, printer and telephone.

AGENT

It is a computer network designed for and by the extension agents, farmers and extension clientele.

- It was first introduced in USA in 1977.
- AGENT is used for problem solving, information sharing communication. Farmers can get answers to their every day management problems through solving function.
- An electronic mailing system serves the communication function by spreading up distribution of message among the staff.
- Accounting function can be greatly facilitated by the use of computer by designing more detailed financial record in less time.

7. Conferencing

Teleconferencing

It is a system for creating dialogue between the geographically dispersed participants. It can make physically distant people talk together.

Audio – conference

The simplest from of teleconferencing is the use of amplified speaker phones to join groups of people at each end of a telephone conference. High quality audio reception and transmission can be achieved. It enables time exchange of information without losing the human touch.

Computer conferencing

In this case, only computer data can be exchanged among multiple locations. It is a variation of e-mail, which enables multiple users, geographically isolated to carryout the dialogue.

In a computer conferencing, participants need not be involved simultaneously. One participant may send information at any time and it may be read by others whenever they have time.

It is basically a group activity. Messages are stored, thus allowing those who join late to retrieve and catch up with conference.

Video conferencing

It is the process of conducting a conference between two or more people at different locations by using computer networks to transmit audio and video data.

Web conferencing

It is defined as the dynamic exchange of all kind of information text. graphics, HTML link to information, audio, video etc in a context of structured conversations organized by item and allowing a participant to contribute spontaneous responses to any item in the conversion.

Example: Making online conversations to promote decision – making, technical support, community building, project management, distance education and electronic meetings etc.

8. Interactive Video

It is combination of computer assisted instruction and video technology. It is an important development in instructional media. It is a computer – mediated system that enables the users to interact with and control information stored on videodisc.

Interactive video is most commonly used for training purpose. This technology is particularly useful in personalizing instruction by allowing learners to proceed at their own pace and receive immediate feedback. The teacher is provided with continuous information on student's performance as well. The concept of interactive video has its roots in individual instruction.

9. VSAT

VSAT stands for Very Small Aperture Satellite Terminal. It is a small earth station for satellite transmission that handles upto 56 kb/sec. Of digital transmission.

10. IGNOU

Indira Gandhi National Open University was started in 1985. It is a distance education council which provide academic guidelines to promote excellence, encourage use of innovative technologies and approaches, enable convergence of all systems and sharing of resources through collaborative networking for access to sustainable education, skill up graduation and training to all.

The university has 48 regional centre, and 5 sub regional centers.

Aim

1. To provide cost effective and quality education to large sections of population including those living in remote and far flung areas.
2. Encourage State Govt. or conventional universities to establish open university or distance education institutes and create a network of such institution.
3. Financial supports, grants and academic guidelines to institutions of open and distance education.

11. HAM Radio

- HAM stands for Hertz Armstrong Radio
- It is also called as Amateur Radio
- Amarendra Chandra Gupta first operated HAM
- Used only in emergencies like calamities and earthquake.
- HAM is operated in Amateur Radio Operation Centre (AROC).
- Code used here is called as Morse Code.

- It has operator and receiver.
- HAM has also been derived from three scientists who discovered it namely Hyman, Almay and Murray.

Education through electronic

Manpower is a tremendous resource but untrained manpower is not in demand. People can be more productive through proper Education and training. Social education implies a comprehensive program of community uplift through community action. Social education, therefore, comprises literacy with emphasis on health, recreation and home life of adults, training in citizenship and guidance in improving economic efficiency. Electronic media offers numerous opportunities for education.

Any human endeavour needs to be planned and managed if it has to have a reasonable chance of success. Thus it is all the more imperative that project managers bring in formal planning and apply management techniques to software projects which build the backbone of the organization's information network.

One could safely proclaim that the internet, mobile personal communication technologies, advanced voice recognition technologies, networking advance. smart-card technological and Linux were amongst the hottest sciences at this year's event.

To become successful in today's competitive world one has to manage the future. Managing the future is achieved by managing the information. Interest is a vast source of information. It is the window to the information super highway.

Internet is the world's largest computer network. It was created nearly 25 years ago as a project for the U.S. Department of Department of Defence. Its goal was to create a method for widely separated computers to transfer data efficiently even in the event of a nuclear attack. From a handful of users in the 1960s today the Internet has grown to thousands of regional networks that can connect millions of users. The Internet is not owned by any single individual, company of country

Internet is a group of two or more networks that are

(i) Interconnected

(ii) Capable of communication and sharing data with each other.

(iii) Able to act together as a single network.

MULTIMEDIA

Multimedia – is a sub field of computer science has progressed to the point that some of its innovative methods are of practical use for information retrieval system design. **Multimedia technology** brings together video, graphics, animation, text and sound in single computer controlled presentation.

Features of a multimedia system

Audio: It is asset of sequence data generated from an oral recording. It is the 'front end' of a multimedia system. It comprises of an audio adapter, an audio input device such as a microphones, software to digitize the analog sound, amplifiers and speaker.

Video: Video is asset of temporally sequenced photographic data. Using the video capturing software via-capturing board, one can convert video sources from television video camera to digital video and save them in popular format such as quick times moves or AVI for PC applications. The standard for digital video is MPEG. Real time video delivery that is delivery the presentations 'live' requires a high hand width network. Today, real time video feels such as those used by the cable. Mews network can be viewed in the desktop.

Animation: It is yet another feature multimedia capability. It can be set of object, two dimensional (2-D) animations is the most however, three dimensional (3-D) animations has mostly been confined to the engineering field. As 3-D animation technology matures and becomes competitive, this feature will get in to applications like on – line tutorials, simulations morphing and virtual reality.

Texts: The existing text information can be available in both printed and electronic forms. Words and symbols in any form.

Spoken or written is the most common system of communication. Cortical voice and sound could guide the audience, but greater effort is required to pay attention to spoken word than to browse text.

Images: Images include pictures and photographs. The storage representation of images is a direct translation of images pixel by pixel. It is stored by using some standard encoding scheme such as JPEG, JPEG which is the standard adopted by the joint photographic experts group for compression and decompression of digital images and more recently video scanned images. Images often need to be processed and enhanced by the use of software like Adobe Photoshop.

Video conference: A way to talk with people and see them on a computer simultaneously, this new technology require, special cameras, monitors, softwares, time, money and patience. In several situations, real time video delivery may not be necessary. In such cases a store forward technology is adopted, in which a user can access a technology of video presentations from his multimedia computer.

Tools and accessories

- **Drawing Software:** There are many hardware and software accessories to create graphics. To create graphics paint or drawing program is needed. These programs can create the graphics ranging from simple squares and circles to complex images. Software available are Corel Draw, Adobe Photoshop, Paint brush on MS Office etc.
- **Scanning:** Another way to create graphics is by scanner. A scanner scans over an image such as photo, drawing logo etc., converting it into an image can be seen on the screen. Using image editors, we can edit image i.e., adding removing colours, filtering etc., can be done.
- **Non-liner editing:** Importing of audio its integration with video is a very simple feature of these soft wares. The audio of two channels and the background and foreground audio can be mixed. The software available is Adobe premiers, media suite pro, etc.
- **Authoring:** On-linear editing with the capability of interactivity is called authoring. Authoring is a tool by which we will able to interactively give any graphics or screens for the user according to his choice. The software available is macromedia director, multimedia tool book from Asymmetries etc.

LATEST TECHNOLOGIES

- **Motion capturing systems:** This is a system by which the motion of various living things are captured and integrated in to the computer and assigned to the computer generated models so that they also behave in a similar manner. This is achieved by fixing sensors all over the body and allowing the person and the object to act accordingly whereby different cameras are used to capture the motion from all sides.

- **3D rapid digitizer:** These are the laser controlled scanners by which the object can be scanned and the output rather than being a 2D image is a 3D Mode. All the effects that are applicable to 3D modeling and animation can be performed to this object.
- **Motion tracking cameras:** These provide a very flexible way of film making. These cameras record their movement and motions and all parameters of the shoot. Later when a computer generated shot is composited on the real shot, the cameras values can be clear synchronization between the computers generated shoot and real shoot.
- **Voice recognition system:** Multimedia is revolutionizing the way we interact with the computers. Voice recognition system enables the computer to recognize commands on voice input using digital format. The first step towards this has already been taken by the department of software like Dragon Naturally Speaking and IBM via voice.
- **Virtual reality:** Virtual reality is a revolutionary facile attribute of the advanced computers of today to envelop a user in a 3D simulated world of sight, sound, touch and movement. There are two classes of virtual reality system, immersive and desktop. The most common tool used on the internet in terms of 3D graphics is VRML: it is a language used to create 3D multimedia tools, VRML is an acronym for the virtual reality modeling language. VRML is a 3D analog to HTML. This means that VRML, serves as a simple, multiplication language for publishing 3D web pages. This is motivated by the fact that some information is best experienced three dimensionally, such as games, engineering and scientific visualizations, educational experience and architecture.

Typically these types project require intensive interaction, animation and user participation and exploring beyond what is capable with a page, text or image based format. VRML provides the technology that integrates three dimensions text and multimedia in to a coherent model. When these media types are combined with scripting languages applications are possible. Many have speculated that the 3D model will supersede and thus replace the popular 2D desktop model as the primary user interface paradigm in the next decade. Of course there are variety of challenges that need to be overcome this is possible such as 3D user interface and navigating and 3D graphics performance.

APPLICATION OF MULTIMEDIA IN AGRICULTURAL EXTENSION (Devaraj, 2001)

Multimedia technology gives low cost per farmers in making contacts, present a viable alternatives. The use of multimedia as a training tool is advantageous to turn agriculture in to industry and enhance turnover, reducing training costs, improving morels of an employee and increase the efficiency and effectiveness of training programmes and reducing the cost of delivery. It is an individual or self learning program.

Following are some examples that illustrate how the multimedia technology function the field of agricultural extension

- An information retrieval multimedia system for extension can be used as primary medium for information delivery. A Multimedia user interface provides a variety pf capabilites including computer graphics and high resolutions digital images. The fully operational multimedia system is being planned to implement in all stages extension offices of the Country.
- The colleges of agricultural sciences are striving to incorporate incorporative innovative electronic technologies, research and extension programmes. The benefit of technologies such as electronic information retrieval and delivery system and multimedia frequently involve dispute about costs, responsibilities and priorities.
- A multimedia computer packages, DIAGONISIS, provides a training aid to extension/farmer for pest and disease diagnosis. The program simulates field and laboratory condition in which farmers must actively seek clues and interpret observations on the cause of plant pathology problems. Output of the programme consist text, graphics and video.
- Interactive multimedia CD-ROM's are most useful for training of extension workers in the production of audio – visual material. Ina computer based multimedia, presentation combined with instructional resources meet the audience's need and multimedia helps to author software programmes. A range of multimedia technologies, includ-

ing text, sound, diagrams, pictures, video clippings, animation and cartoons can be used to present the subject matter in an interesting, interactive and effective format, photographs, picture etc., of symptoms of disease, insect pest damage and of insect can also be made.

Multimedia application in other areas

- Entertainment
- Broadcasting
- Advertising
- Marketing
- World wide web
- Publishing etc.

Interactive videodisc

An Emerging Technology for Educators. ERIC Digest.

Students in English 10 have been directed to write an essay about the culture and major events during the time of John Steinbeck's classic GRAPES OF WRATH. They have also been given a computer – based interactive video system with which to search for relevant information. Controlled by a computer, a major database of textual information is available through CD-ROM and computer disc; in addition, a database of visual and audio information is accessible through a videodisc. During their eager search through the information, the students run into problems with one of the concepts and press "lesson" to be tutored on the topic. The tutorial uses both text and visuals to make important points. For a final small group of the issues of the day. the unit concludes when the teacher uses the interactive video system in front of the class to summarize features of the exercise. Video and text are at teacher's fingertips with the simple press of a few buttons. Discussions are lively because the experience has enticed the students by immersing them in the sights and sounds of earlier has enticed the students by immersing them in the sights and sounds of earlier times.

INTERACTIVE VIDEO?

As demonstrated in the above scenario, interactive Video can be a very complex learning system, or it can be a simple tool for teachers to use to enhance their instruction. The term has been used broadly in the literature and includes three major aspects: 1) interactive video as storage, 2) interactive video as hardware, and 3) interactive video as learning concept. the purpose of this digest is to describe each of these concepts in detail.

INTERACTIVE VIDEO AS STORAGE DEVICE

A videodisc storage unit is a 12-or 8-inch disc that looks a large version of an audio compact disc. read by a laser beam is not affected by fingerprints. Videodisc has two formats, each offering different capabilities : CAV (Constant Angular Velocity) and CLV (Constant Liner Velocity). In addition, there are two tracks that can store a total of 30 minutes of random access audio. On a CAV disc, information is stored in tiny pits pressed in concentric rings sealed between protective layers of plastic. Information in each ring can be assessed in less than three seconds. Besides compact storage, a major advantage of interactive video is this rapid precise access to any bit information on the disc.

A CLV disc spins at a constant linear velocity, much like a long ply record. Because of this, CLV discs can hold much more data than CAV discs: up to one hour of straight play motion video for a 12-inch discs can hold much more data than CAV discs. up to one hour of straight play motion video for a 12-inch disc and 20 minutes for an 8-inch disc. This format allows random access through time code rather than frame number. Because time access is not as precise as frame access, the major advantage of a CLV disc is its compact storage capability.

INTERACTIVE VIDEO AS HARDWARE

Interactive video hardware is needed to run a videodisc. The Nebraska Design/production Group developed a widely adopted classification. Scheme which groups the hardware by "levels of intelligence" (Daynes & Buter, 1984). The higher the level; the greater the ability for internal computer control. Contrary to expectation, the more control the system has, the easier it is for the learner to operate and ignore the hardware.

- **Level 0.** This describes a stand videodisc player that is used for simple linear playback. Many movies are stored on CLV discs and are played in this fashion.
- **Level 1.** A Level 2 player is also a stand-alone videodisc player, but it has additional built-in functions for rapid and precise frame searching, variable motion,mono or

stereo audio control, stable freeze frames, and scanning, A level I player allows the teacher to display any information in the visual and audio database to an entire class. A Level I Player can also be used by students, individually or in small groups, with a keypad and printed directions.

- **Level 2.** A Level 2 player has all the functions of a Level 1 player plus a built-in microprocessor and user-programmable memory. This type of player is most useful when programming is encode right on the second audio track of the videodisc. Level 2 players allow more internal control through this programming on the disc. they are often seen in smalls, information booths, or science displays where the customer is invited to press simple, One-key response. While this system hs its greatest use outside the classroom, it can be used by the teacher for individualized or small group enrichment exercises.
- **Level 3.** A Level 3 player has all the capabilities of a Level player and is connected to a microcomputer. Current combinations of videodisc players and computer system offer a choice of one or two screens. In a two screen system, one displays the video while the other displays the computer text. In a one screen system, video and computer text are displayed on the same screen. A one screen system, with an appropriate overlay board, Often the hardware comes with a variety of components for interaction, such as a touch sensitive screen, mouse, or light pen. Because of the computer "Command center," Level 3 interactive video offers the most flexibility for use as a large group , small or individual learning device.

Since the development of this classification scheme, some have defined another level. Level 4 has come to mean any level 3 player with additional hardware such as a second (or third) videodisc player, a CD-ROM drive, or other electronic equipment.

INTERACTIVE VIDEO AS LEARNING CONCEPT

Fleming and Levie (1978) classify between instruction and learning according to the locus of the activity instruction occurs outside the learner, but learning is something only the learner can do. As shown in the opening scenario, interactive video can be a true learning device since it is not something that is done to the learner but, rather, is an activity three learner does himself or herself. Wittrock (1989) states that in order for learning to occur, learners must be, at a minimum mentally, active in the learning process. All computer-based instructional system can mentally enage learners by directly involving them in learning. But because of the added dimension that video offers, interactive video surpasses other computer-based system in its ability to involve the learner and engage him or her in a two-way dialog.

Well designed interactive video lessons invite learners to interact with the visuals by enabling them to interrupt, identify, sequence, and select from alternative actions. Students can interrupt visual scenes when they identify problems in the scenario. they can stop sequences if they can't follow the ligic or get lost. they can practic with concepts by matching pictures to words, identifying critical visual elements of concepts, or matching sounds to words. For sequencing activities can select approprite psychpmotor sequences or select options can be shown consequences visually and dramatically (Ippa and Anderson, 1988.)

Interactive video has been used in almost all content areas from literature to mathematics and science. The literature is full of case accounts of its effectiveness for a variety of uses. It most be remembered, however, that it is not the medium which makes interactive video effective, but the way in which it has been implemented.

Iuupa and Anderson (1988) state that interactive video is "more that a new trick, a new device or even a new technology" (p.139). Interactive video as storage and hardware blends the elements once available concept creates an environments where the 1989). Interactive videos as learning concept creates an environment where the two-way dialog essential for learning has become a reality. Because costs are falling and research has demonstrated the importance of interactivity, interactive video is destined to play an important role in the future of education and training.

1. The government machinery is not able and prepared to bear the burden of expenditure on public extension system and privatization of extension services is being encouraged without any restriction and assigned role and responsibilities.

This is creating a misleading situation and confusion. the fund, which is divested through private players, needs to be restricted to a considerable extent and be utilized to strengthen the CD extension system with clear-cut responsibilities and accountability.

2. In theory the private extension is more efficient and effective, but in reality, it is not. It is more selective to justify its presence ignoring the most needy people. The work done by public extension system has not been recognized and now the system is being questioned for everything. The extension staff have never been rewarded or given incentives and even their positions remain static since the beginning. A critical look is needed on these to revamp the public extension system.

3. The private extension works on the specific areas and can meet the demand for the specific information more efficiently than public extension system because of its network. Therefore, the private services need to be properly channeled and utilized to support the main CD system. particularly for post harvest management, Entrepreneurship development though small-scale cottage infrastructure and export promotion.

4. ICT diffusion and infusion in agriculture provide necessary digital opportunities or advantages for productivity increase, income generation, decrease in national disparity and improving linkages with the market. Grassroots level information access (contents) and grassroots level access to information (networking) are the two essential components ICTs (a framework) for sustainable development and liveihoods is essential at grassroots level (at society level)

5. Net working of people and networking of information are required for reaching the unreached public extension system. Preparation for future development and communication among masses require an extensive thinking regarding information technologies and convergence with mass media including the internet. An important development is the emergence of e-Commerce. More and more companies are setting virtual show-rooms, which can be assessed by an users from any corner of the world. Internet and the facility of world wide web (WWW) offer unique opportunity of near communication. Such support of private extension is required to make the public system more vibrant.

6. The challenging need is to pack all technological information into digital media based web presentations so as to be used in the ICT families created for distant learning through village knowledge centers. information kiosks, digital media forums, face to face learning through CD, video conferencing and web-based presentations. However, this cannot become substitute players are helpful in the matter to support the public extension system.

7. Several research studies conducted on extension organizations have revealed that the delivery of goods is effective when the grassroots extension worker covers a small area of jurisdiction, with multiple purposes. The existing system of large jurisdictions, each with a narrow range of activities, is less effective. The broad basing requires grassroots extension workers to be at the cutting edge of extension and master of many trades. IT strengthening of public system can help enable extension staff to gather, store , retrieve and disseminate a broad range of information needed by farmers, thus transforming them from extension workers into knowledge workers to realize much talked about bottom-up, demand driven technology generation, assessment, refinement and transfer. ICT can help the extension system in re-orienting itself, towards such agricultural development and also play an important role to document both organic and traditional cultivation practices in form Traditional Knowledge Digital Library (TKDI) to be shared more widely.

Strengthening ICT for sustainable development

- Dissemination of sustainable agricultural package through different modes of ICD
- Developing CDs related to sustainable developments which include relevant case studies
- Providing two way interaction between different development agencies and stake holders through video conferencing
- Training on computer literacy is required to educate the rural mass about the usage of ICT
- Establishment of portals to link other leading websites like FAO, CGIAR etc
- through this networking establish forward linkages with teriary industries like value addition, processing, etc.
- SAUs, Central Research Institutions and leading industries will act as source of knowledge for ICT
- local language and content should blend with ICT for better acceptance by the farming community.

Successful Cases of ICT enabled services

Computer aided extension services: Operating in the present knowledge based era, speedy dissemination and acquisition of information holds the key for success. The farmers must keep themselves abreast with the latest information about their enterprise to sustain their feet in the market. The researchers, the extension agencies and development of commodity specify Decision Support System (DSS) and Expert system. On-line market info-portal could bring about improvement in precision decision making by the growers and lead to betterment in production and trade of horticultural produce.

Chapter – 9

Information and Communication Technologies

Cyber Extension

Information and Communication Technologies (ICTs) such as the world wide web, e-mail, telephones, fibre optics and satellites are revolutionizing the way in which societies interact, conduct their businesses and to compete in international market. It also contributes towards strengthening democracy, increasing social participation, competing in the global market.

About 69 per cent of US citizens are surfing the Net. African population has 3 per cent access to internet. In Asia, we have 10 per cent internet connectivity. China (8.5%) has taken the lead over India (4.5%) in terms of internet connectivity, Japan (68%), South Korea (67%), Singapore (67%), and Taiwan (60%) are far ahead.

Recent experiments show that ICTs can have very positive impact among the poor. The innovative usage of the mobile phone resulted in alternative means of communication, empowerment of Muslim women and the mobility of poor people in the village. The info-village programme of M.S. Swaminathan Foundation in Pondicherry is able to deliver daily images obtained from a web site run by the US Navy of the predicted wave conditions in the Bay of Bengal at Veerampattinam. The villagers who are fisher folk are able to learn the sea conditions and decide whether or not to go for fishing on a given day. This information that is so crucial for their lives, is transmitted verbally across a public address system, from loud speakers, to these fishermen as they prepare their boats in the early morning.

The impact of ICTs is witnessed in many aspects of the day-to-day lives of the poor. ICTs can have a major role in reducing the impact of natural disasters on the poor in low-income countries. The World Ban's hazard management programme in high risk areas of Andhra Pradesh, involves ICT components in cyclone warning, communication and response, awareness raising, education and community involvement in hazard reduction activities.

Although there are many positive effects, there are also many constrains faced by governments, institutions and individual seeking to introduce these tools for enhancing livelihood opportunities. Constraints such as policy failures, lack of private financing, lack of quality content, poverty and illiteracy, poor infrastructure and so on make it very difficult to access them or to use them as effectively as desired. It is widely acknowledged that the influence of these technologies may affect the way in which enterprises operate, make-decisions, produce or exchange intangible goods and services such as artistic creation, learning, knowledge sharing.

Cyber Space is the imaginary or virtual space of computers connected with each other on Networks, across the globe. These computers can access information in form of Text, Graphics, audio, video and animation files. Software tools on networks provide facilities to interactively access the information from connected servers. The Cyber space thus, can be defined as the imaginary space behind the interconnected telecommunications and computer networks, the virtual world. Agricultural Extension, according to Dr. D. Benor, "relates to the process of carrying the technology of scientific agriculture to the farmer in order to enable him to utilize the knowledge and a better economy". Cyber Extension thus, can be defined as the Extension over Cyber Space. Cyber Extension means, "using the power on online networks, computer communications and digital interactive multimedia to facilitate dissemination of agricultural technology".

Cyber Extension includes effective use of Information and Communication Technology, national and international information Networks, Internet, Expert Systems, Multimedia Learning Systems and Computer based training systems to improve information access to the Farmers, Extension Workers, Research Scientists and Extension Managers.

Elements of Cyber Extension

E-Mail, On-Line Extension Information Services, Interactive Expert Systems on crop pests and diseases, INTERNET browsing for Extension Information, Video Conferencing-Static, Mobile and Call Centres, SATCOM Net works are major elements of Cyber Extension.

Technology Cluster

- The Wireless in Local Loop (WILL)" The Technology which can provide Rural Connectivity in most accessible areas – already tested in over dozen locations in India.
- Simputer and Nicetop: (low cost computer and communication devices developed indigenously)
- Intranet (the Gyandoot Model): Server based application providing e-Governance interface at village level.
- Mobile Video conferencing: Using Satellite Communication (SATCOM) technology.
- Digital Video Broadcasting: Extended C-band KU-band based digital broadcasting (both data and video)
- Radio, TV and Call Centres: Integration of Media options.

Information Kiosks: Current Outreach

Total number of KIOSKS are around 26000 and 4.33% of villages in the small country are connected. Corporate sector has established 57 per cent of the KIOSKS. Average investment on one Information Kiosk is about Rs. 65,830. One Kiosk serves upto 225 farmers. The farmers in the age group of 35-40 were the highest in number to seek information from Kiosk.

KVKs and ICT

A huge network of KVKs have been developed by the ICAR in India. Out of 589 rural dustricts, 536 rural districts have been covered by establishing a KVK in each district. There are 8 Zonal co-ordination units in 8 zones of ICAR in the country to monitor, review, facilitate, develop action plans, and provide funds to KVKs in their respective zones.

Review of KVK system indicates that there is limited technological backstopping of KVKs from research institutions. SAUs and other such organizations, support in terms of internet connectivity is very selective. There is an emerging scope realized by the Indian Council of Agricultural Research to bring all the KVKs in a network where all the KVKs are electronically connected with all the technical institutions. For this purpose, V-SAT connectivity is to be provided to some selected KVKs. About 10 subject matter specialists/technical experts are posted at each KVK, which makes a pool of 5360 well trained experts/scientists in the country. Such a treasure of human resource can be utilized more effectively by providing ICT network/VSAT connectivity at KVK level. Networking established by any private/public organization at village level may be used by KVKs for linkages with the farmers. In this way, most of the villages can be covered through ICT connectivity.

There is lack of reliable telecom infrastructure in rural areas. Erratic or no power supply is another concern. There is lack of ICT trained manpower in rural areas. No relevant content is available in local Language. Lack of information services to rural clientele is faced during performing the service. Low purchasing power of the rural communities; lack of holistic approach and issues of sustainability are the other major issues and concerns.

Conscious effort is required at policy level to integrate ICT enabled extension as a part of agricultural extension. There is necessity of a nodal planning and implementation agency at country/province level. Continuous hunt and integration of new communication technologies is required in Cyber Extension agency. Use of Media-Radio and TV is needed in participatory programming and interactive mode. Efforts to promote the use of call centers for Agricultural Extension need to be popularized. There is need of continuous monitoring and updating of all extension systems and databases. Mechanism of capturing lessons and sharing horizontally, vertically is to be created.

Information Village Project

The project is implemented by M.S. Swaminathan Research Foundation, Chennai in collaboration with IRDC, Canada.

Value Addition Centre (VAC)

Hub of the Information and Communication network is established at Villianur – a major Market centre, administrative node a road junction. Village Information Shops (VISs) are established at Kizhur, Embalam, Veerampattinam, Nettapakkam and Pillyarkuppam.

Internet access through VSNL. Reading room (Books and Documents in Tamil on aspects of Agricultural production;

Collection of Government notifications); Library of Educational CDs for learning; Information Access.

Databases maintained at VAC include families below poverty line, entitlements to rural families (schemes for the benefit of disadvantaged groups), local transport information, directory of health specialists, price related information for both produce, and inputs (fertilizer and seeds) PASIC provides this info on a daily basis on women and child health, health programmes and Directory of Health Specialists, environmental protection, micro-meteorological info relating to the local area, medicinal plants, price information for both produce and inputs, market related information, transport, IPM database for paddy, sugarcane, vermin-composting, etc.

Village Information Shops

Shop located at a place was identified by the community (PRA exercises carried out to identify location, volunteers, meeting running costs, charging for information etc.) Volunteers Operating Shops were identified by the community. Information Group was formed with 1:1 ratio for men and women to guide the functioning of the VIS. Another aspect is contrast between the roles of ICTs at different levels of society. For rural users info revolution is not the open democratic phenomenon as they a accessing info assembled and packaged in specific formats. They have no alternate sources. Despite language barriers and mediated databases there are grounds for optimism. The Information Revolution is bearing fruit.

ITC e-Choupals

'e-Choupals' makes use of the physical transmission capabilities of current intermediaries – aggregation, logistics, counter-party risk and bridge financing – while disintermediating them from the chain of information flow and market signals. Village internet kiosks managed by farmers – called Sachalaks – themselves, enable the agricultural community access ready information in their local language on the weather & market prices, disseminate knowledge on scientific farm practices & risk management, facilitate the sale of farm inputs and purchase farm produce from the farmers' doorsteps.

Real – time information and customized knowledge provided by 'e-Choupal' enhance the ability of farmers to take decisions and align their farm output with market demand and secure quality and productivity. The aggregation of the demand for farm inputs from individual farmers gives them access to high quality inputs from established and reputed manufacturers at fair prices. As a direct marketing channel, virtually linked to the 'mandi' system for price discovery, 'e-Choupal' eliminates wasteful intermediation and multiple handling. Thereby it significantly reduces transaction costs. Launched in June, 2000, 'e-Choupal', has already become the largest initiative among all Internet-based interventions in rural India, 'e-Choupal' services today reach out to more than 3.5 million farmers growing a range of crops – soybean, coffee, wheat, rice, pulses, shrimp – in over 36,000 villages through nearly 6000 kiosks across nine states (Madhya Pradesh, Haryana, Uttaranchal, Karnataka, Andhra Pradesh, Uttar Pradesh, Maharashtra, Rajasthan and Kerala).

This model uses proprietary network of intermediaries for procurement and distribution 5,050 kiosks in 31,000 villages are spread.

TATA Kisan Sansar

The TATA Kisan Sansar network runs cop clinics where agronomists use computers to access information from the geographic information system (GIS) and advises farmers on what to grow, where and when to grow it, and how much urea and nutrients to use. At the soil testing laboratory, technicians analyse soil samples to determine their composition and confirm what the satellite maps have indicated. Additionally, the TKS network operates experimental farms where scientists conduct agricultural research and development. The command area of the Tata Kisan Sansars covers Uttar Pradesh, Haryana and Punjab. The network of farmer centres is divided into mother TKSs known as Tata Krishi Vikas Kendras (TKVK) and franchisee TKSs. Each TKVK covers 20 franchisee centres, and each franchisee about 60 villages. Currently, 40 TKVKs and about 800 franchisee TKSs are in operation, catering to 27,200 villages and almost 25 lakh farmers.

I-Kisan Limied

I-Kisan Limited is an initiative of the Nagarjuna group of companies, established in 2000 with the objective of facilitating continued enhancement of agricultural productivity and rural prosperity, and thereby making the Indian farmers globally competitive. Initiative includes both I-Kisan portal and Informaion Kiosks at the block/village level for disseminating information in Hindi and five major regional languages (Tamil, Telugu, Kannada, Marathi and Punjabi). The portal has organized information on agricultural practices, agricultural news, animal husbandry, agricultural machinery, aromatic and medicinal plants, agricultural credit, insurance and prices of inputs. It also available at each kiosk for disseminating information and providing services of I-Kisan. A `member registration' system has been evolved for I-Kisan at the village level with the twin aims of developing a rapport between the farmers and I-Kisan and stimulating revenue generation through specific service charges. This initiative has benefited the farmers in terms of crop yield increase (5%) and reduction in cost of cultivation (14%) mainly due to the judicious use of farm inputs.

Each initiative is a unique model in the application of ICTs to agriculture and has merits and constraints of its own. To sum up, for effective utilization of ICT, policies need to be focused on (a) Media planning for need based content development and mode of delivery; evolving capacity building programs for facilitators to improve their effective interfacing, (b) Devising strategies to integrate quality services and needed infrastructure to support ICT-enabled services, (c) Optimizing partnership among public and private institutions and NGOs to perform complementary roles in ICT-led growth, (d) Incentives for use of advanced and alternative technologies of ICTs to overcome regional barriers (remote and under developed) in connectivity and (e) Increasing public investment for rural infrastructure development and launching explicit ICT-based agricultural development programmes. For building ICT network of KVKs, a holistic view of HRD, infrastructure development, establishing forward and backward linkages and motivation is required.

Ref: Singh, A.K, Lakhan Singh and R. Roy Burman, 2006, Building ICT Net work for KVKs, National Seminal on Information and Communication Technology, Opportunities and Challenges for Revitalizing Extension System, Indian Society of Extension Education, December 27-29, Navsari.

Information Technology Center

Six Information Technology Centres have been established to give training to staff, students and farmers of the village. These centers also function as computer booths and are maintained by a booth operator. The Centres have been provided with Six Pentium II computers with configuration of Pentium II, 64-Mb RAM, 2GB hard disk and a dot matrix printer. A dial up connectivity with modem and telephone line has been used to connect the central server station to retrieve information, send the queries, grievances to the main center.

NIC, Pune has developed a Computer Based Question Bank, in the local language "Marathi" which will be used to test the Computer knowledge awareness of the students of 5th to 10th standard. These students are being trained to get acquainted with the computer systems. Testing will be in subjects covered in the school like mathematics, science etc. and a certain percentage of marks will be awarded from this test to the final marks of the students. A batch consisting of five students will be examined for 1 hour. This center also serves as a computer booth. NIIT is engaged in helping create CDs on different topics which when available will be used at these centres for interactive coaching.

The Team from MANAGE which carried out the study also had the opportunity to interact with farmers, field operators. The field officer Mr. Kulkarni, Agriculture Officer, Panhala village at Kolhapur district expressed that Cyber Extension would be the major form of technology dissemination in the near future and added that the advantages of the WAN systems were on spot data availability for different purposes, avoidance of unnecessary movement of farmers from their home village to data centre for different purposes and access to all the information at village level.

An interaction with a few farmers in villages of Paragoam, Kuralap, Bhariwadi etc. indicated that farmers like the concept. They believe that the information from Wired Computers (WAN) is major source of getting information on crop technology. The ranking given by the farmers for source of information on crop technology, ranks wired computers as the best source followed by field officers and staff, radio and TV, print media and company sales persons. They agree that WAN provides all necessary information on cultivation technology and market situation etc., information is timely, the language is understandable as it is in the local Marathi language, information is reliable and it is not costly. They also agree that it is beneficial not only for big farmers but also for marginal and small farmers. The farmers are expecting some more features to be added like processing of all loan, legal documents etc. from their village wired computers. The farmers are also ready to learn operating computers, if any training is given and are ready to contribute any marginal amount necessary for the maintenance and up keep of the system.

Access to information and improved communication is a crucial requirement for sustainable agricultural development. Modern communication technologies when applied to conditions in rural areas can help improve communication, increase participation, disseminates information and share knowledge and skills. It is being said that "Cyber Extension" would be the major form of technology dissemination in the near future.

However, it is observed that the rural population still has difficulty to accessing crucial information in order to make timely decisions. It is essential that information availability is demand driven rather than supply driven. The Challenge is not only to improve the accessibility of communication technology to the rural population but also to improve its relevance to local development.

Technology in Rural Communication

A range of communication technologies have been used to meet the information needs of the rural population. These include rural radio, participatory video using a methodology called Rural Audiovisual pedagogy which uses participatory video as a communication tool for mediating between rural peoples needs and possible sources of information and expertise to respond to these needs.

The Internet is emerging as a tool with potential to contribute to rural development. Internet enables rural communities to receive information and assistance from other development organizations, offers opportunities for two-way communication. It can also support bottom up articulation of development needs and perceptions, and thus help in reducing the isolation of rural communities. It can facilitate dialogue among communities and with government planners, development agencies, researchers, and technical experts encourage community perception in decision making; and help agricultural researchers, technicians, farmers and others in sharing information.

A survey of Internet use in rural areas was conducted by Dr. Don Richardson and others in 1996. Rural Internet users indicate that the Internet provides them with a very convenient method for quickly accessing a large volume of information without being impeded by geographic barriers. They also report finding information of value from the Internet in the form of new ideas, discussion groups, access to expert advice, continuing education resources, increased global understanding and cultural awareness. Social benefits include new opportunities to overcome geographic isolation, increased social interaction, opportunities to organize and advocate for social change, euqualization of urban/ rural disparities and new links between urban and rural communities.

Information is an important resource in modern agriculture. The development of computers and improvements in telecommunications offers farmers and extension workers, many new opportunities to obtain technical and economic information quickly and use it effectively for their decision making. The modern farmer is an entrepreneur who tries to grow right crops and animals in the most profitable way. The amount of information a farmer can and should use for his management decision is increasing rapidly.

CYBER EXTENSION - ADVANTAGES, LIMITATION AND APPLICATION

Cyber extension is defined as the extension over cyber space' cyber extension means using the power of online networks computer communication and digital inter active multimedia to facilitate dissemination of agricultural technology. Cyber extension includes effective use of information and communication technology, national and international information networks, Internet, Expert systems, multimedia learning systems and computer based, training systems to improve information access to the farmers, extension workers, research scientists and extension managers.

Advantages

Cyber extension is a tool that gives access to the vast global information resource.

- It enables rural communities to receive information and assistance from development organizations and offer opportunities for a two way communication
- The dialogue among communities and with planners, development agencies, researchers and technical experts can kike place through the internet and also encourages community participation in decision making and helps agricultural researchers, extension managers, technical experts, farmers and others in sharing information.

- On the spot data availability for different purposes, avoidance of unnecessary movement of farmers from their homes or villages to the data centre for different purposes and access to all inform-ation at village level.
- It is essential that information availability is demand driven rather than supply driven.
- Cyber extension and its application offer enormous potential in enhancing the efficiency of implementation, monitoring and evaluation of various development programs of the agricultural research education extension, planning of crop production,
- It also acts as a powerful medium for the transfer of modern technology to the farmers.
- Farmers can also benefit directly through the utilization of various websites that have been launched by the government generally, the collection of information on various aspects of crops.
- Cyber extension is a global encyclopedia on virtually any topic - very technical to very retrivial.
- Accessible 24 hr pcr day. When ever the farmer has time or the inclination no need for making a meeting appointment.
- Technical level of information - ranges from general Public information to press release extension information - research report production information.
- Authentically/Accuracy of information there is no gover-ning internet group that regulates the integrity of the displayed information.
- Cyber extension when implemented through community information centre can facilitate extension worker to deli-ver demand driven technologies to be farming community.
- The web is an ideal publicity and trading medium which is why large number organizations have been putting up web pages.
- Internet websites are another tool for the farmer to acquire information and solution.
- It saves time
- Information is rich and interactive
- Instant international reach

- Continuously available.
- Electronic version of messages, no printing, updated messages.
- User friendly, easier to access
- Linking experts' vs. farmers.

Limitations

Cyber extension is becoming an tool for development communication, however most rural communities are not yet able to take advantage of this new tool due to illiteracy, lack of skills, economic power.

There will be both irrelevant and relevant messages.

Cyber extension is more expensive.

- It cut steps from extension process like trainings and zonal workshops.
- All messages will not be suitable to all farmers.

The methods of cyber extension

1. Providing interaction among research scientists, extension workers, farmers and other rural people through e-mail
2. Providing up to date news and information services such as market prices and weather conditions.
3. A question and answer service where experts respond to queries on specialized subjects
4. Creation and maintenance of statistical databases on critical agricultural and rural development parameters that can be decried on demand.
5. Providing the details of poverty alleviation schemes on the Internet.
6. Providing status of various government programmes and details about their implementation mechanism on demand basis.
7. Hosting websites by major institutions participating in agricultural extension, putting latest packages of practices for various agro-eco regions. These institutions, particularly the project Directorates may also

place the diagnostic and pre-emptive farm practices for the major crops particularly the commercial crops, well in advance of the concerned crop season. This can help the extension workers to access latest information on IPM, INM and other such practices for high value important commercial crops. The institutions will also be able to get direct customer feedback for their packages.

8. Launching online rural development and extension journals, newsletters etc.
9. Providing internet access at district and block level agriculture and rural development offices. This service may be open for rural communities on fixed days. This connectivity can also be used to download online publications on useful topics from anywhere in the world.
10. Opening of cyber cafes to enable educated rural people and extension workers at village level to have direct access to world wide web having market information etc.
11. Providing maps that display different features, such as population density, crops planted, etc.
12. Providing video clips to demonstrate complex procedures and audio files for rebroadcast on local radio stations.
13. Providing mechanism of user/beneficiary feed back for the public sector schemes.

Application

For scientific community

- Document delivery services
- Date services
- An aid in analysis of research programme
- A central source of information of (he nature, location *and* status of current research project
- Planning implementation of Agricultural research information systems.
- Generation, analysis, integration arid Diffusion of national and international agricultural research information

- Maintenance of agricultural information system and Participation in the international information system on Agricultural Sciences and Technology (AGRIS) of FAO of the united Notions.

For extension personnel

- Bibiliographic services
- Document delivery services
- Technical enquiry services
- Immediate and accurate information services
- Extensive and efficient compilation of results from different branches of agriculture

For policy makers, administrators and media personnel

- Consolidated information (regarding practical use of technology)
- Accurate and efficient data services

For farmers

- Conventional library and information services Catering mostly to extension literature
- Efficient collaboration with scientific community in solving their day to day problems
- Farm management services
- Data recording and maintaining services
- Medium of sharing their local information with each other.

Information or advice on the following

Both the current and future prices in the market for various alternative crops. Weather forecasts for his area that would decide or alter his cropping timetable. The availability of fertilizer and pesticide inputs their vendors and prices. Where to get the seeds for his next crop and at the lowest price from vendors across the country and even overseas. Immediate information on forecasting and diagnosing pest and disease attacks and when and where to market his produce.

Chapter – 10

Modes of Conferencing

EXPERT SYSTEM

An Expert System is designed to simulate one aspect of a human expert, the ability to classify phenomena from a set of attributes. The expert system is a classification engine. It is a system that takes information about a particular case or stance within the domain of the system and produces a qualitative result. It usually has [Complete information and makes qualitative judgments based on this information.

Expert systems, as a class of computer programs, are currently designed to reflect general model current within the artificial intelligence community; an expert system is not simply a simulation of human expertise, but must be implemented (on a computer) in a particular fashion; it is a product of an AI culture. Ideally an expert system has two primary components viz. the knowledge base and the inference engine.

What is expert system model?

Expert system model comprises of structure of the internal parts of the expert tern through which an expert system is developed and tested for its effectiveness in Id conditions.

Here, some of the models are given to create an expert system.

STRUCTURE

The internal structure of an expert system can be considered to consist of threes: the knowledge base, the database and the rule interpreter.

The knowledge base holds the set of rules of inference that are used in reasoning. Most of these systems use IF-THEN rules to represent knowledge. Typically systems can have from a few hundred to a few thousand rules.

The database gives the context of the problem domain and is generally considered be a set of useful facts. These are the facts that satisfy the condition part of the condition action rules as the IF THEN rules can be thought of .

The rule interpreter is often known as an inference engine and controls the knowledge base using the set of facts to produce even more facts. Communication with system is ideally provided by a natural language interface. This enables a user to jet independently of the expert with the intelligent system.

Knowledge Base

The Knowledge Base is essentially a set of rules describing relations between elements in the domain of knowledge. The rules for deriving an outcome from a set of additions are always formulated externally by an expert usually aided by a knowledge engineer, i.e. a specialist in transforming the expert's information into statements suitable a knowledge base. The knowledge that is selected for inclusion in the knowledge se can have a variety of forms, depending on the form of the inference engine. Most system designers consider it important that the rules be easily inserted, modified, or deleted from the knowledge base, in any order. In nut shell, knowledge base is the Election of information.

Interface

It is the front page. It is the window. It is used to get the advice.

The Inference Engine

The inference engine takes the rules provided by the knowledge base, and uses internal rules of inference to draw a conclusion. The claim is that the internal rules are general to all inference. So the inference engine is a set of rules which arc applied to the les in the knowledge base. The inference engine is intended to be based on a general (del for using knowledge and should not have special knowledge about a particular) main. Expert systems should be viewed as a particular type of information system, systems are distinct in terms of their approach to problem representation, as information system process information, while expert system attempt to process knowledge. Knowledge in an expert system may originate from many sources, such as textbooks, reports, databases, case studies, empirical data, and personal experience. The dominant source of knowledge in today's expert systems is the dominant expert. A knowledge engineer usually obtains knowledge through direct interaction with the expert. In the expert systems there are widely used in many application areas. There are few types of problems solving paradigm such as control, design, diagnosis, instruction, interpretation, monitoring, planning, prediction, prescription, selection and simulation.

Knowledge engineering methodology is given here with explanation.

What is an Expert System?

An expert system is a computer program designed to simulate the problem solving behavior of a human who is an expert in a narrow domain or discipline. An expert system is normally composed of a knowledge base (information, heuristics, etc.), inference engine (analyzes the knowledge base), and the end user interface (accepting inputs, generating outputs). The path that leads to the development of expert systems is different from that of conventional programming techniques. The concepts for expert development come from the subject domain of artificial intelligence (AI), and lire a departure from conventional computing practices and programming techniques. conventional program consists of an algorithmic process to reach a specific result. An I program is made up of a knowledge base and a procedure to infer an answer. Expert terns are capable of delivering quantitative information, much of which has been eloped through basic and applied research (e.g. economic thresholds, crop development models, pest population models) as well as heuristics to interpret qualitatively derived values, or for use in lieu of quantitative information. Another feature that these systems can address imprecise and incomplete data through the assignment confidence values to inputs and conclusions.

Components of an Expert System

All expert systems are composed of several basic components: a user interface, database, a knowledge base, and an inference mechanism. Moreover, expert system development usually proceeds through several phases including problem selection, knowledge acquisition, knowledge representation, programming, testing and evaluation.

Knowledge in the Expert System

The knowledge the expert uses to solve a problem must be represented in a don that can be used to code into the computer and then be available for decision making by the expert system. There are various formal methods for representing knowledge and usually the characteristics of a particular problem will determine the appropriate representation techniques employed.

Knowledge bases can be represented by production rules. These rules consist of a condition or premise followed by an action or conclusion (IF condition...THEN action).

Explanation of the Decision Making Process

One of the attractive features of expert systems is the program's ability to review a consultation and provide the user with an explanation for how its conclusion was derived. The explanation function is essentially a record of the reasoning process used by the expert to resolve the problem. It provides for a better understanding of how the conclusion was reached and instills in the user a greater confidence in the conclusion and expert system. The accumulation of facts to be presented when an explanation is asked for is usually part of the development shell or software.

Making the Expert System Easy to use

Whether or not an expert system achieves success may be determined by the nature of its user interface. This is the part of the expert system that interacts with the user. Even the most powerful expert system will not be applied if it requires too much effort on the part of the user. For this reason, if is important to make the computer as easy the user to operate as possible. Almost all modern development programs offer capacity to interact with the system through both text and graphics.

Developing an Expert System

The power of an expert system is derived from the knowledge of the expert. The 8c of computer skills should not inhibit anyone from using this tool in implementing jar disease management program. There are published procedures that one can follow develop expert systems, although rigid adherence to these procedures is not necessarily prerequisite for a successful development effort.

Selecting the Problem

The development of expert system is discussed in three parts:

- Selecting a problem
- Knowledge acquisition and representation
- Evaluation and adoption.

The most critical step in developing an expert system is identifying a suitable problem. Expert systems are best suited to problems that require experience, knowledge, judgment, and complex interactions to arrive at a solution.

One of the first tests to determine if a subject area is suitable for an expert system is ether the solution of the problem requires the knowledge and expertise of a human expert. In addition, in crop areas where little is known about certain diseases the power of expert system is really not necessary.

Cost of Development

Expert systems are expensive to develop. They require resources, expertise and to build. It should be determined if the proposed expert system can be justified in of savings or other benefits it produces. Costs include software, hardware and personnel to do the work. A developer should ask if the problem to be solved justifies the investment of time and money.

Knowledge Acquisition and Representation

After the problem has been selected, the knowledge acquisition phase of expert tern development is begun. The task now is to have the knowledge which the expert to solve the problem displayed in a logical fashion so that it can be coded into the computer.

Most expert system developers suggest that a knowledge engineer (someone who is lined in the extraction of information from experts) extract the knowledge and design fee expert system. In some cases, a knowledge engineer is needed to 'tease apart' highly compiled expert knowledge because the expert is too familiar with the area to break it into its logical components. There is also software available to help in self-knowledge engineering.

Steps followed in designing an expert system

The user interface is created by drawing controls such as text box, labels, option I button, command button etc., on a form. Each form is carrying single question of a cause or about a situation coupled with dichotomous options of Yes or No. AI form is one of the main building blocks in this application. Users interact with the controls on a form to obtain the desired result. The form becomes the first on screen window a user sees while running a programme. In order to make the programme I easier to read and debug appropriate names are assigned to the objectives.

The properties are set for the form and controls to specify values such as the caption, colour, size etc. Codes are written to bring the application to life or after setting the initial properties for the form and each object, codes are added that executes in response to events.

HTML template-An Hypertext Markup Language (HTML) page is designed to act as a template. Some parts of the pages are replaced with dynamic content before the page is displayed especially, advice and more details of the cause are designed to act as a template.

Describing symptoms in words is very difficult and sometimes is very confusing. Therefore, necessary images are inserted into relevant forms, with the main purpose of describing a disorder symptom and confirming the diagnosis of the cause of a certain disorder by the user.

The voice is recorded and edited using a software sound forge. It is one of the best sound editing and recording software and the background sound is given to add flavour to the multimedia presentation.

After the project has been created, it is saved using the 'Saved Project As' menu item on the 'File' menu option. Saving a project also saves its form and code modules.

Debugging tools are used to check for errors and modify code whenever necessary. Debugging means taking over the corre-ctions. It is like pre testing.

Empirical model of an expert system *a*

Balasubramaniam (2003) developed empirical model of an expert system rubber crop and tested the effectiveness of expert system. Here the pest problems in rubber crop are selected. The user friendly software ESS shell software or VB6 is used. The diagnostic path started from the origin of pests to how it ends up as an endemic pest is included.

Expert Systems of other Countries

Knowledge representation and expert system tools have been applied successfully into Agriculture in China, such as adequate fertilization, plant protection cultivation water resource Management and irrigation, raising and management of livestock and poultry. The Expert system has pushed the growth of Agriculture and also the research of Artificial Intelligence in China. They have produced evident economic effect and social impact.

Expert Systems for Cotton Crop Management

Expert System has been developed by the US Department of Agriculture provide appropriate management recommendations to cotton growers.

CALEX

This is a blackboard based integrated expert decision support system for agricultural management, developed at University of California. CALEX can be used by growers, pest control supervisors, consultants and other managers.

VARIEX

This expert system developed at Technical University of Brno, Czechoslovakia enables selects of the best cultivators for different agricultural situations.

Wiping Jin Expert System

There are fertilizing, irrigating, spraying insecticide process and adopting other measures in crop management, which rely on crop state analysis. However, there is a lot of fuzzy phenomenan in crop state analyse. Then it is necessary to analysis or to measure these fuzzy phenomena by means of fuzzy set theory for crop management. An expert system that can deal with fuzzine been developed for crop management in PR China.

Strengths of expert system

1. Practical experience and knowledge of human expert can be stored and used for longer period of time
2. Helps for quick decision making
3. Self learning is possible
4. Highly convinced about diagnostic result
5. Highly interactive
6. Saves money
7. Save time
8. Reduces the dependency of human expert

Weakness of expert system

1. Required computer another accessions
2. Needs regular uptake whenever situation changes
3. Skilled personnel required to uptake the knowledge base of the expert system
4. Computer literary is required operate
5. Useful for only literate farmers
6. Need power supply
7. Language barrier
8. Low credibility in the diagnostic result

Opportunities in using expert system

- Steady growth of information communication would help to use the expert system
- Declining trend in the cost of computer and other accessions
- Possible to bridge the gap between scientist knowledge and farmers
- Widespread computer literary noticed among children's of farmers in rural area
- Literary level of Indian farmers is at an increasing pace
- Potential to develop expert system in local languages
- Solar power/battery operated back ups can be used to run the computer system

- Possible to update the knowledge base of the expert system

Threats in using expert system

- Wrong knowledge gathering and representation lead to wrong conclusion
- Cant' rely upon the ICT, sometimes makes operational problems
- Viruses cause corruptions and hang over of system

Survey of Expert System in Agriculture

The works carried out on Expert System in agriculture and allied field and various software used to develop an Expert System by authors were collected worldwide and presented as follows.

S.N.	Authors	Name of ES	Utility	Software/Shell used
1	Lemmon (1986)	COMAX	ES for cotton crop management	-
2	Palmer (1986)	COMAX	Soybean crop variety selection	PROLOG
3	Shroyer *et al.* (1987)	WHEAT WIZ	Cultivator selection tool	-
4	Bennett and Sneed (1988)	COMAX	Planning, design and evaluation of irrigation systems	PASCAL
5	Floris *et al.* (1988)	COMAX	Real-time operation; real-time metcorological data handling	PASCAL
6	Getforth and Macvieer (1988)	OASIS	Operation of control structures; real-time meteorological data handling	PASCAL
7	Haic and Irwin (1988)	EXSYS	Drainage diagnosis	PASCAL
8	Halterman *et al.* (1988)	ES	Double cropping management	-
9	Boggess *et al.* (1989)	FinsARS	Financial analysis for farm business management	-

10	Stone and Toman (1989)	COT FLEX	Cotton crop management; coupled with SOYGRO model	PASCAL
11	Batchclor *et al.* (1989)	SMART SOY	Soybean crop management	Insight 2+
12	McClendon *et al.* (1989)	SMART SOY-IRRIG	Soybean irrigation	Insight 2+
13	Morgan *et al.* (1989)	CUE	Crop variety selection	SELECT
14	Hart *et al.* (1989)	CUE	Irrigation system selection	LISP
15	Hershacur *et al.* (1989)	CUE	Canal water distribution; canal network incorporated	LISP
16	Bhatty (1990)	RESEXP	Reservoir operation; DP model integrated	PROLOG
17	Helms *et al.* (1990)	CIRMAN	Crop insurance strategies	-
18	McGregor and Thornton (1990)	CVSES	Wheat crop variety selection	CRYSTAL
19	Oswald (1990)	TANK	Tank systems diagnostic analysis	PROLOG
20	Han *et al.* (1990)	ES	Sprayer diagnostics	-
21	Hasbini *et al.* (1991)	PUMP	Operational guidelines for center pivot systems	PASCAL
22	King et al. (1991)	MKBS	Fertilizer & irrigation applications	Turbo C
23	Mevo and Amir (1991)	CROPLOT	Multiple crop selection	Rabbi
24	Srinivasan *et al.* (1991)	ESIM	Delivery system operation; canal network incorporated	EXSYS
25	Clake *et al.* (1992)	IRRIGATOR	Irrigation scheduling; ET method selection	PC PLUS

26	Elange *et al.* (1992)	BDM-EXPERT	Drought management integrated with CASIMBOL model	IITM RULE
27	Kumar *et al.* (1993)	KBS	Economic feasibility of irrigation system selection	Level 5
28	Nakamura and Tsukiyama (1992)	ES	Irrigation canal renovation project planning	-
29	Plant *et al.* (1992)	CALEX/cotton	Cotton irrigation scheduling	CALEX
30	Raman *et al.* (1992)	BDM-EXPERT	Crop planning under droughts; LP model inferencing	Insight 2+
31	Bralts *et al.* (1993)	ES	Hydrologic analysis of micro irrigation system	-
32	Mohan and Arumugam (1994)	CROPES	Multiple crop selection	IITM RULE
33	Nevo *et al.* (1994)	CROPLAN	Optimal crop planning; LP model integrated	PROLOG
34	Pasqual (1994)	ES	Identification and control of weeds in wheat, barely and oats	-
35	Arumugam (1995)	TANKES	Tanks system operational guidelines; real-time operation	VP-EXPERT
36	Mohan and Arumugan (1995)	ETES	ET estimation method selection	VP-EXPERT
37	Nuthall and Bishop- Hurley (1996)	-	ES for animal feeding management	VP-EXPERT

The studies reviewed under this section clearly indicates that, various software's were used by the researchers to develop computer based Expert System and used as an effective tool in various fields of agriculture. The above observations suggests the need to develop a user friendly computer based Expert System considering the flexibility, simplicity, nature of problem and familiarity of the f software to the student researcher.

Experiences in using Expert System for Agricultural Development

Bundy (1984) stated that the application of Expert System generally falls under three classes, namely, Expert System proper, intelligent front ends, and hybrid systems. An Expert System proper is a purely rule based system, relying on a sizable knowledge base. It is based on a qualitative, causal understanding of how things work. Such a system is more suitable under situation wherein not quantitative data are used. It is essentially conceptual and heuristic rule based system. An intelligent front-I end is a user friendly interface to a software package, enables the user to interact with the computer using his/her terminology. It minimizes or avoids misuse of complex models by Jess experienced users. A hybrid system represents the integration of algorithmic techniques with Expert System concepts.

Lindsey *et al.* (1989) opined that the recent developments in information dissemination techniques like a CD-ROM technology, and Expert Systems can improve the efficiency and effectiveness of international agricultural research in developing countries, especially for crop protection against pests.

Morgan *et al.* (1989) described on expert System for crop variety selection. They developed a system which was designed to consider the soil characteristics, water availability and prevalence of diseases This system was developed for winter wheat in Scotland. This would allow agricultural extension officers to recommend varieties with confidence thereby reducing the demand for advice from specialist crop advisors.

Spiangler *et at.* (1 9S9) observed that an expert's knowledge is the central and key component of developing an Expert System. Furthermore, it is more difficult component. At the end, the knowledge acquisition must be regarded as much as an art as a methodical and scientific procedure. One approach, however, that often seems to be ignored is the collection, integration and use of research results, while an expert should clearly build research results into their expertise, it is also possible to bypass the expert and use the published results in formulating rules in cases where the research provides a complete and logical answer.

Bhatty (1990) presented a hybrid Expert System for optimal operation of a reservoir system in Pakistan. This Expert System includes the cognitive and computational components involved in the reservoir operation. The reservoir operation has the objectives of irrigation and hydropower generation. This Expert System was mainly intended to maximize power production.

Depradine (2003) introduced, CITOR, Code Information Extractor, an Expert System designed to extract syntax and structural information from partially complete or incorrect Java code. The information can then be used by development tools to provide various checks in real-time.

Ghosh and Samanta (2003) developed and tested a rule-based, object oriented Expert System for insect pest management in tea code named TEAPEST. The system identifies major insect pests of tea and suggests appropriate control measures. TEAPEST shows good performance as evident from its performance evaluation.

Jayawardhana *et al.* (2003) developed a user friendly Expert System BESTCOMP: Expert System for Sri Lankan solid waste composting for better management of solid waste composting by local authorities in Sri Lanka. BESTCOMP Expert System mainly focused on the behaviour of the physical, chemical and biological process in composting. The intention had been to provide distant users with scientific and techno economic information using modern tools but at a much lower cost.

This research has put very strong emphasis on allowing the user to browse around the knowledge that has being extracted from books, published research articles, reports, audio, video, Internet, case studies and the domain experts who involved in solid waste management activities, so the user can get an accurate and a real feel for the solid waste management subject.

Oswald (1990) has studied the feasibility of using an Expert System for the tank diagnostic analysis. Diagnostic analysis is usually performed by experts who try to find the reasons for the malfunctioning of tank irrigation system and also to identify possible remedial measures. The limitation of the study is that the basic knowledge is derived from only two tank irrigation systems in south India. Hence, the general utility of this Expert System to other tank irrigation system is limited.

Schiefer and Christensen (1990) stated that the broader utilization of information technology in farm management activities is primarily restricted by a lack of appropriate concepts. The development of such concepts traditionally involve a functional formalization of the flow and the processing of information, In addition, recent advances in artificial intelligence research have opened new visits for concepts which are based on human reasoning (Expert Systems) and which could complement functional forms to allow die formulation of more comprehensive concepts and consequently, a further integration of information technology in farm management activities.

Batchelor *et al.* (1991) revealed that Expert Systems methodology has shown considerable promise as an information technology. However, limited knowledge of how current information technologies relate to the decision process impedes the adoption of Expert Systems. The significance of developing an economic theory of Expert Systems is substantiated with an empirical application investigating a soybean pest management decision process (SMARTSOY) based on experience with four insect pests causing damage to soybeans in the southeastern USA. SMARTSOY is combined with SOYGRO (soybean crop growth simulation model). Pest management recommendations from extension bulletins and the expert system are compared with an expert's recommendations. Results indicate the potential improvement in decision making processes with die adoption of Expert Systems.

MULTIMEDIA

Multimedia - concept

It is an application that brings together multiple type of media like text, illustrations photos, sounds, voice, animations and video. A combination of three or more of these with some measures of user interactivity is usually referred as multimedia.

- The power and promise of multimedia gives control to the end user.
- Users can navigate and get relevant information. At present scenario information grows more precious, multimedia is becoming one of the most effective way of getting that information.
- A picture is worth a thousand words, a full motion animation with voice over narration and back ground music stress the importance of multimedia.

Multimedia Authoring Tool / authoring software

- It is the application software used for integration of the elements such as text graphics animation, sound, video etc., into a presentation and incorporation of refractivity through navigational control is called multimedia software or simply authoring tool.

- It brings the different elements of multimedia in to a comprehensive presentation.
- It also incorporates interactive elements such as navigation tool into the presentation.

Functions of Authoring tools

The multimedia authoring tools have four basic functions:

1. Importing: The tool should have capability to import the building blacks such as text, graphics, animation, audio files, digitized video files and other software.

2. Creation: The authoring tool should have inbuilt capacity to generate or create the building blacks by designing, editing, painting, charting, audio capturing and video processing.

3. Integration: It should enable integration of the building blocks by sequencing, timing, linking and scripting. It should also capable of keeping database for user responses, provide feedback and call on other resources and application to be run from the multimedia application produced.

4. Delivery: The authoring tool should be capable of developing self running application and encryption of the application.

Type of Multimedia Authoring Tools

It can be classified based on the way by which the software organize the elements sequences the events and deliver the application, *i.e.* based on the metaphor on which the tool is based on it can be classified in to four types.

Time based

Card / page based

Icon Based

Theoretical concept based

1. Time based: It organize and present the multimedia elements along a time - **line with well** defined information on when the elements are to be played at what speed and for how much duration they have to be played.

- This type of tool is very useful when the elements have to precisely synchronized.

- Movement through the element is only sequential or linear.

2. Card/ page based: In this system, elements are arranged in relational database as stacks of cards or books of pages, DVR to relationally one can move randomly from randomly from any card/ page to any other card/page in any stack/book.

The limitation of linear movement as in case of time based systems is overcome by developing interrelationships b/w the card/pages

Users can jump/ branch to the desired card/ page.

Such a system allows the use of navigation buton, background layer common to all the cards / pages

3. Icon Based: Here, the elements are iconized and the authoring tools allow visual programming for sequential as well as logical flow of events by dragging icons into the programming abandons from the icon menu.

One of he most powerful icon based authoring tools is the macromedia Author ware professional.

4. Theoretical concept based: It consider the building blocks as CAST and SPRITES that appear on STAGE (screen) at defined point of time, for a defined period and behave in the way directed by the attached behavior SCRIPTS. The elements are created or imported and stored as a database in the cast window, these casts are moved into the score window where they are sequenced, animated, timed find, liked. This can produce elaborate and complex stage shows.

- Macromedia Director is a very popular and Powerful authoring tool based on the theatrical metaphor other way of classification based on the capability and complexity

1. Multimedia capable Tools: These tools put together the elements in a sequential or linear order. So the interactively in the production as using these tools in almost

These tools are basically intended for non - professional applications mostly for education or training purposes.

2. Dedicated media integration Tools: There tools are not capable of generating the elements. They simply import the pre - generated building blocks and arrange the elements in the time sequence needed. Thus are dedicated only for integration of elements. These tools are used only when all the elements have been generated with other software applications.

3. Professional multi media Tools: This tool offer built in programming capabilities and an ability to generate various media elements with in the program itself. These are very powerful but highly complex tools used for developing highly interactive application such as instructional software, training packages and games. Due to its complexity only professionals utilize these tools. Interactive multimedia authoring tools such tools have capability of developing and integrating interactively at any level complexity. The goal of interactivity is to provide the user with some feed back and control over the applications It allows the user to choose a particular content. These liable integration of interactivity by organizing the elements in several layers of formation and inter relate them through navigation elements.

The navigation tools provide links among various sections of the elements and follow any of the following four structures,

1. **Linear structure:** Allows for sequential/ linear movement through the elements by the users.
2. **Hierarchical structure:** The user can navigate through a branching structure defined by the logic flow of the application.
3. **Non-linear structure:** The navigation through the application is possible without a pre-determined or prescribed path.
4. **Composite structure:** The user can be navigate freely through the presentation, except for certain locations where the logic sequence of the application decides the navigation.

Authoring tools for specific purpose

Proposed used Proposed used	**Authoring Tools**
Education	Macromedia Director, Author ware, Asymetrix Tool Book HyperCard, Hyper studies, Super card.
Business	Macromedia Author ware, Action, A stound
Games	Macromedia Director, HyperCard super card
Cyber Application	Macromedia Director, Dream weaver Microsoft front page, Adele page mill.

MULTIMEDIA APPLICATIONS

Multimedia, which encompasses all the traditional forms of communication, likes video stills, graphics, audio, animation and text is an excellent tool for extension communication. The computer integrates all these media on to a single platform and provides interactivity to the system. This is essentially the power of multimedia, integrating varied media forms and giving control to the learners to let them use it exactly the way they want to do. Most popular multimedia technologies are Interactive Video Disc (IVD) are Digital multimedia.

The use of multimedia as training tools is advantageous to turn animal husbandry and agriculture into industry and enhance turnover, reducing training costs, improving morale of an employee and increase the efficiency and effectiveness of training programmes and reducing the cost of delivery. An information retrieval multimedia system for veterinary/agricultural extension can be used as primary medium for information delivery (Devaraj, 2001).

Advantage of multimedia

To impart knowledge for a group oil trainers learners or individuals and creating awareness of subjects/technologies/ innovation.

To impart skill on a subject matter to individual or group of farmers.

To get convinced to get in-depth understanding and to teach illiterate farmers. For any technology to be discriminated among farmers, it should be transmitted in a convincing way to the Farmers.

The presence of combination of audio, video, symbols and their application simultaneously through hyperlink and action button would help farmers to get in depth understanding and to teach illiterate farmers.

AGRI-PORTAL

Agri-Portal acts as a gateway for farmers to find information they seek in the agriculture and rural domain. Agri portals are established as an electronic network. The information's on the technology packages are made possible to the farmers through these portals. Portal sites provide gateways to evaluated agricultural websites ID to find answers to enquiries.

It is the time to promote portals for farmers for bringing technological improvements. In these heady days of "e-everything" agriculture should not go far behind. Agri portals are the logical next step in our four decades of progress in the modernization of agriculture. As computers attract more, users, agriculture is going to be online in a big way. All this will be done in a mouse click away.

Portal

Portal is a gate way, for a World Wide Web (WWW) site. Typical sites offered by portal sites include a directory of web sites, a facility to search/link other sites and other details like specific information on any choosen topic, news, weather, mail, block quotes, chat facilities among people, all these at the click of mouse Srinivasan (2002).

Agri portal provides services not only to satisfy one type of customer but all stake holders involved in agriculture. By accessing in to portal different types of services can be enjoyed. Portal provides an elaborative information including micro details.

Web site

A web site is a related collection of World Wide Web files that beginning file called a home page. A company or an individual tells how to get their web site by giving address of their home page. From the home page, one can get to all the other pages on their site. Transaction is not possible in web sites. Only information exchange can be done. The information contained is also not static Srinivasan (2002).

There are three types of web sites based on the IP address used. They are, .ac.in - used for academic institutions, e.g.: www.tnau.ac.in .org.in - used for organizations, e.g.: www.fao.org. .com - commercial or business purpose, e.g.: ikisan.com

Dotcom

A dot com is a web site intended for business. The term is based on the 'com' that forms the last part of the address for most or the commercial web sites.

www has a fire wall protection (i.e.) security. Registration should be done to create a web site. Only one web site will be available with the particular name and capacity is also given. No one can middle and only the creators have access to the of contents for alteration. Then it is taken to the web server. Any search it is the time to promote portals for farmers for bringing technological improvements. In these heady days of "e-everything" agriculture should not go far behind. Agri portals are the logical next step in our four decades of progress in the modernization of agriculture. As computers attract more users, agriculture is going to be online in a big way. All this will be done in a mouse click away.

LIST OF AGRI PORTALS

Agri portals related to India

1. Krishiworld.com: India's First Multilingual Agricultural Portal. The comprehensive interactive site that gives the latest farming oriented news and information.
2. IndiaGgronet.com: site dedicated to the development of agriculture and the farming community.
3. Kisanmdii.com : Multilingual Portal providing agricultural information
4. Kissan.com: An agricultural portal providing basic information needs of farmers.
5. Agrisurf.com: An extensive agricultural information portal operating globally.
6. Agriwatch.com: An agricultural portal providing customized agricultural commodities market analysis reports.

SITES RELATED TO AGRICULTURE

A. Government of India related

1. Department of Agriculture and Cooperation: http://agricoop.nic.in
2. Department of Fertilizers (Directory of Links) http://fert.nic.in http://goidirectory.nic.in
3. Agriculture Markets: http://www.agmarknet.nic.in
4. Indian Meteorological Department: http://www.imd.gov.in
5. National Centre for Medium Range Weather Forecasting http://www.hcmrw.gov.in
6. National Informatics Centre: http://www.nic.in

7. Fertilizer Association of India (FAI): http://www.faidelhi.org
8. International Fertilizer Industry Association: http://www.fertilizer.org
9. Food and Agricultural Organization (UN): nttp://www.fao.org
10. Indian Council of Agricultural Research (ICAR): http://www.icar.org.in
11. Indian Agricultural Research Institute (IAR1): http://www.iaripusa.org
12. Indian Societies of Agribusiness Professionals: www.isapindia.org

KAP is a network of agriculture and allied sector professionals in India and the SAARC countries. It is supported by Dept. of Science and Technology, ICICI and Ford foundation.

Services provider by some important portals:

Market Information's

Agriwatch brings to daily online reports from various agriculture markets in India. Reports from each of these markets cover daily details on prices, arrivals, stocks etc. Commodity wise spot market prices are announced

http://www.agricultureinformation.com

This portal provides information's related to agri services, crop details, education research, farm inputs, organizations, processing industry and other useful resources. For all the services link information is also given.

Agri Services

It includes agri tourism, auctions, certification, commodity trading, consultancy, contract farming, e-commerce and employment details.

Crops

Link information is available for the crops includes cereals, fibre crops, field crops, floriculture, forestry, gardening, grassland and pastures, horticulture, legume crops and medicinal plants and herbs.

Education and research

Directory of services is extended for aquaculture and fisheries, agri business, agri extension, agricultural economics, agricultural engineering, agronomy, animal biology, animal science, bio science and bioinformatics.

Farm inputs

Informations related to all sorts of inputs *viz.* biological inputs, crop nutrients, crop protection, farm equipment, farm technology, planting materials, etc. is given in link.

Livestock

Livestock information's like aquaculture, animal feed and supplements, animal its welfare, beef cattle, buffalo (bison), dairy cattle, donkeys and deer and emu farming is also available in this portal.

Organisations

One can get the informations related to organizations viz. cooperatives, government, international organisations, museums, non profit organisations, political parties, professional associations, super markets and trade bodies through the services provided in this portal.

Processing industry

Directory of services is also available for processing industries like additives, baby food, baked food, beverages, brokers, canned food, condiments and seasonings, confectionary, dairy products and de-hydrated items.

Useful resources

Other useful informations provided in this portal are books, conferences, consumer information, databases, discussion forums, journals, magazines, market news, market research and newspapers.

Other services offered on this portal at present are:

1. Multi search Engine and Web Directory

This search engine and web directory could be used to identify interesting agriculture websites. The advanced meta search engine of this portal which simultaneously searches all major search engines. It is possible for anyone to search all the agri-business websites in one place instead of having to visit several separate search engines.

2. Business Promotion Services

This portal offers many interesting ways to generate sales for products and services and helps thousands of businesses everyday in locating potential customers for their agri products and services around the world.

3. Download Centre

Download catalogues and brochures of agriculture companies.

4. Bulletin Board

It is an another form of e-mail. This is the place to post messages regarding any queries on agriculture. It meets other agri entrepreneurs and growers and exchange ideas. Problems can be solved by experts in the discussion forum. Any body looking in to the bulletin board could get the comprehensive service.

5. World Agri News Centre

Latest articles and news reports on agriculture compiled from the world's leading newspapers and magazines could be read. They update these reports throughout the day one can even customize the news centre to only display headlines in the area of interest.

6. One can sign up for the free web mail service and get the yourname@agri123.com email ID for agribusiness. This email ID is very popular with growers, agriculture businesses, agriculture university students etc,

7. We designing studio

Using the web design studio one can build a low cost and attractive website. The web site can be used to sell products and **services** online and reach out to new overseas buyers.

Services of special interest to Indian audience

a. Virtual Library

It has the collection of highly readable and useful articles on new business opportunities in Indian agriculture. One can sign up for a membership and access a wealth of high quality commercial agribusiness information that is not easily found elsewhere.

b. India Agribusiness Newsletter

It pertains to business opportunities in Indian agribusiness.

c. Vadamalai Consultancy Scrvices (VCS)

VCS helps in preparing the hi tech farming project and assisting many traditional farmers in India to take up cultivation of high value medicinal crops.

d. Ikisan

Ikisan is being developed as a comprehensive agri portal to address the information, knowledge and business requirements of various players in the agri arena viz., farmers, trade channel partners and agri input/output companies, leveraging information technology and extensive field presence, ikisan is being positioned as an information/knowledge exchange and an e-marketplace. The services provided in this portal are given below.

Agricultural information's

Crop Specific informations and informations related to animal Husbandry, agricultural machinery, annal credit, insurance input and output trading, market information's, weather forecasts, commodity news, product availability, online loan facilities, chat rooms and discussion forums etc. are available in this portal.

Market information

Ikisan provides commodity prices, commodity arrivals and analysis reports for major commodities at local market level on daily basis.

Partners

To provide complete value to the entire agricultural community Ikisan has tied with major companies and service providers.

News

Keep in touch with the world at large is the key requirement for the farmers. Also specific news on agriculture - the developments, there in market, movements are of immense importance. Ikisan has tied up with leading news providers, for the latest news and updates on happenings in the agri world and the world at large.

Weather

Weather plays a major role in agriculture. Ikisan provides weather data and also satellite images of the forecast.

Local information

This contains sponsored pages related to agromet advisory, district wise seasonal rainfall, rainfall deviation, crop wise area sown.

Doing business

To enable companies' interactions with the trade channel, Ikisan has developed Ikisan vyapar and Ikisan mandi as communication and exchange platforms for the agri input and agri out piri companies respectively.

KISAN BAZAR is the best place to buy or sell agri produce. It is an open market for all kind of agri Produce. It gives the chance to list items for sale or purchase in the specific groups like seeds, planting material, fruits, vegetables, flowers, medicinal and aromatic plants, cereals, oil seeds, spices and organically grown produce.

Applications / Benefits of agri portals to the farmers

- The real motivator for the adoption of the computer by farmers will be ease of access to information, low cost of trade and access to information, low cost of trade and access to more goods and services at lower
- Access to latest agricultural technologies, market trends, news, weather, subsidies and relevant information's.
- The farmers can in time, appreciate the benefits of seeing for himself the prices in different parts of the country, thus avoiding exploitation by middlemen.
- Access to on line chat with consultant experts
- Advisory services on variety of farming needs and sharing of experiences with other farmers
- The farmers can log on to the site of fertilizer, seeds, pesticide providers, nurseries, etc. and order directly, saving money.

- The farmers can sell their produce to the highest bidder online and simultaneously book a transporter to cart it away.
- Directory of services and service providers like soil testing and seed testing laboratories can be availed.
- Products such as the kisan card will give farmers the credit and resources, he needs to trade online.
- Wastage and exploitation will be eliminated to a great extent.

Constraints in using agri portals

- Non availability of larger number of telephone connections in rural areas.
- Illiteracy of farmers may be a major barrier in availing the facilities provided by agri portals.
- Phobia to using of computers.
- Occurrence of frequent power failures in rural areas.

VIDEO CONFERENCING

The power of computers and the flow of data double every 18 months. Networks are beginning support the convergence of voice, data and video applications. These sophisticated applications and breakthrough technologies continue to improve our daily lives. With the faster processing and the affordability and usability of the Internet, this great invention is now changing lives and the meaning of communication. Whether separated by an office building, a department, a city a time zone or even an ocean, video conferencing technology enables people to connect. This technology is simple to use, yet highly functional and offers flexibility and the interactivity that makes it highly adaptable to curriculum and instructional strategies. The rich and advanced technology 'video conferencing' is discussed here.

Concept and Definitions

In telecommunication, the term Video teleconference has the following meanings:

1. A teleconference that includes video communications.
2. Pertaining to a two-way electronic communications system that permits two or more persons in different

locations to engage in the equivalent of face to face audio and video communications. *Note:* Video teleconferences may be conducted as if all of the participants are in the same room – www.bambooweb.com

"Video conferencing technology allows two or more people at different locations to see and hear each other at the same time. In addition, it is often possible to share computer applications such as Internet pages, library catalogs, documents, or software. This rich communications technology offers new possibilities for schools, college and libraries including formal instruction (courses, lessons, and tutoring), connection with guest speakers experts, multi-school project collaboration, professional activities such as meetings and interviews, and community events" www.kn.pacbell.com

"With integrated video, voice, and data capabilities, video conferencing lets any one connect and collaborate anytime, anywhere. Teach a class, participate in a meeting, or talk one-on-one with a colleague or interview a job candidate with full video and audio enhanced features allows to play DVD or VHS tapes, show a PowerPoint or other application directly from your Laptop, or use a document camera to share information. Video conferencing allows you to communicate with locations around the world with full video, high-quality audio and computer files (such as presentations)"

"Video conferencing in its most basic form is the transmission of image (video) and speech (audio) back and forth between two or more physically separate locations. This is accomplished through the use of cameras (to capture and send video from your local endpoint), video displays (to display video received from remote endpoints), microphones (to capture and send audio from your local endpoint), & speakers (to play audio received form remote endpoints)"

History of Video Conferencing

Prior to video, few methods of long distance live communications were available. Audio conferencing, such as telephones and on-line chat programs allowed for real-time conversations, but did not allow for much in the wave of personal response. Seeing the person being spoken to is an integral part of communication, and could not be satisfied with then current technologies. With the coming of video, options for communication expanded.

- The first public videoconference was held in April 1930, between two cities of New York. It was at this lime that the value of face-to-face conversation at a distance was expressed.

- In 1971, the first transatlantic videoconference occurred.
- And some twenty years later (90's) desktop video conferencing clients became available.
- Intel, Picture Tel, and VTEL, names with which we arc still familiar today, were some of the early desktop players.

Types of Video Conferencing

There are two types of video conferencing.

Point to point conferencing: It is a communication link between any two locations.

Multipoint Conferencing

It is a link between varieties of locations (more than two). Multi point conference can further be classified as two types, LAN and MCU (Multi Control Unit). LAN provide links between different locations groups etc. In a Multi Control Unit environment, audio and video signals are automatically switched during conferencing (perhaps to attend to bandwidth issues).

Desktop video-conferencing (DVC) is an emerging technology of video-conferencing that uses a standard personal computer which has been enhanced with special video processing capabilities and a small camera. This technology is employed by a broad spectrum of users to facilitate cost effective communication with key individuals or groups regardless of their geographic location.

Hard ware Requirements

A videoconference system must have audio-visual equipment (monitor, camera, Microphone, and speaker) as well as a means of transmitting information between sites. A broadband satellite connection with studio-quality equipment produces an excellent full-motion video connection; but expense is huge. Advances in telecommunications technologies and lower costs have sparked an interest in compressed video systems which transmit information via the Internet using Internet Protocols (IP) and digital telephone networks such as ISDN.

The equipment used to conduct video conferencing between two stations is given here and may be increased proportionately.

2 Video conferencing pioneers

2 Video conferencing terminal end stations

2 Peripheral sets, including camera, microphone, and speakers.

2 Workstations

1 High speed network connection (al least 128Kbps.)

2 IP addresses for workstations,

1 Multipoint Conferencing Unit (MCU) - optional.

Application / Benefits

1. Reaching remote places

Video conferencing enables participants to see, hear and communicate with others at remote sites. It has enabled communication to restricted areas such as nuclear facilities, operating rooms and the space shuffle. It is possible for the researchers to take advantage of labs *which* might be thousands of miles from their own location. It increases connection with the outside world. It makes a face to face communication,

2. Education

- Video conferencing allow to have experts in class rooms and to access people and resources that would otherwise be not accessible. It is ideal for teaching every session to students at remote locations, or invite a guest speaker to joint local class for one day
- On line studies : Several studies land support to the idea of on-line studies as being equal in quality to classes involving everyone in the scene room. Reports find that on line teaching methods present results which run average with those of more traditional styles.
- Video conferencing provides students the opportunity to learn by participating in 2-way communication. They receive instruction and information on any topic, and they present information and ask questions of the other participating site.
- Video conferencing provides a channel for communication all over the globe. The benefits of video conferencing include stronger speaking, listening, and Presentation skills for our students. Following is a list of uses for video Conferencing in the classroom.

- Bring a NASA astronaut to classroom; visit a museum, volcanoes, zoo aquarium and virtual field trips - etc.
- Demos on physical education can be conducted.
- Video conferencing for music students for live one-on-one guitar lessons.
- Student projects (of diverse locations) It is very common to assign group projects, particularly in higher level classes and as term projects. This is a good team building strategy which allows the students to tackle larger problems and learn from each other. As long as the students have been located at the same campus, or reasonably close by, this works well. While application sharing and data collaboration could still be used locally (say for those night owls who don't want to drive late at night), a great deal of diversity can be added to the project if the students are in separate locations. Students in environmental studies might be teamed together from diverse locations such as a coastal environment, a mountain environment, a desert environment, etc. The students can use data collaboration to prepare their final reports, run data analysis for all to see, etc.

Heightens Motivation

- A video conference can improve retention and appeal to a variety of learning styles by including diverse media such as video or audio clips, graphics, animations, computer applications. Educators and librarians from around the country reported that videoconferencing technology impacts student learning.
- The excitement of using new technology and interacting with other Students or an adult increases motivation.

Increases depth of Learning

- Students learn to ask better questions
- Learning is from a primary source rather than from a textbook
- Students show more depth in understanding
- Necessary planning contributes to a better learning experiences.

Presentations : such as lectures, seminars, and courses: With higher video bandwidth networks, lectures and seminars would be delivered to participating researchers, lab users, etc. through video conferencing.

Publications: reports, journal papers, conference presentations: Standards based representations for electronic imaging, textual information and graphical information, and a methodology for efficient co-authorship would further the electronic exchange of data and facilitate the preparation and dissemination of reports and publications. Video conferencing could be used for joint preparation.

Distance Learning: Video conferencing provides opportunity for variety of distance learning classes according to the individual needs. Boston University is currently using this technology for the same purpose. Learning foreign languages and cultural exchanges is also possible through video conferencing.

Training

- Specialists can speak from a location to a room full of people.
- Lectures of best teachers can be showed to other teachers to improve teaching skills. It is a good way to mentors new teachers.

Organisations

- Video conferencing can also be used within organizations to provide immediate telepresence using internal LAN as the communication infrastructure.
- Perhaps the most familiar application associated with video conferencing is that of business communications. Previous to video technology, the only means of communicating live over long distance were a telephone call or an on-line chat program. These methods allowed for real-time conversation, but took away the satisfaction of seeing the person being spoken to, and did not allow for the subtleties of body language for aiding in negotiation.
- The advent of video communications made it possible to do all of the above at once. The person on the other end of the conversation is once again made visible, regardless of distance. What is further, parties from multiple other

locations can all join in on a single conference as if they share a room. Multiple persons on one call was possible before video; however, never before had a person in Japan been able to be face-to-face with both a fellow professional in America and another in France simultaneously. The medium for allowing a full-scale board meeting with members from around the world was now available.

- Video-conferencing describes the use of compressed video technology for live, two-way, interactive communication in a variety of situations person to person, informal discussions, formal group meeting, and large lectures. The primary use of video conferencing is to allow the timely exchange of information without traveling. Most meetings that are held face-to-face can be held by video conferencing. Video conferencing is useful in situations of crisis management, or times when meetings are held on short notice.

Telemedicine

- Telemedicine is a rapidly growing field which, in large part, is made possible by the development of videoconferencing technologies. Health care is one of the largest sectors of the American economy, and any way to provide cheaper or better care is in demand. Video conferencing can be used many ways in telemedicine, from inexpensive low end desktop systems to high end room based systems.
- One application currently being used in many rural areas is for remote clinical diagnoses. In this scenario, a rural doctor consults with a doctor in a metropolitan or university hospital. This allows the remote doctor to see the patient, talk with the local doctor, examine data, and in some cases look at radiological data. This application typically requires a high end system with high data rates and traditionally a Telephone line or several ISDN lines. This can be very expensive. For example, several Southern states run their own statewide telemedicine system linking hospitals, rural clinics, prisons, nursing homes and more. These systems have proven to be cost effective, even though the equipment cost per site has traditionally been morethen $100,000. Using new H.323 videoconferencing system may lower the costs significantly for use in this application, allowing for widespread deployment in more rural and urban areas.

- Another emerging application for telemedicine is home health care delivery using a variety of technologies, including videoconferencing. In these applications, a number of patients who are undergoing treatment for chronic conditions might be sent home with a PC based telemedicine system. This system will connect back to the hospital or doctor's office via a high speed connection such as ISDN. ADSL or a Cable Modem. The system will combine various medical diagnostic devices, such as pulse oximetry, ECG, or Doppler ultrasound with a video conferencing system. This allows the doctor or nurse to examine the patient remotely and assists them in conducting medical readings and gathering the data for the central site. Another possibility is to put a terminal in the doctor's home so that he/she might be able to quickly provide assistance without needing to travel into the hospital. This provides high quality of life to the patient while reducing the costs involved in treating the patient.

Telecommuting

- Telecommuting is a fast growing trend in many large urban areas, as companies and employees realize the many benefits to be gained by this. Video conferencing can provide a natural, productivity enhancing improvement to telecommuting and can bridge the gap between employees who are at home and those in the office. Currently video conference to the home can be done using ISDN lines. However, the home use of video conferencing will not become wide spread until increased bandwidth to the home is in place, using such technologies as cable modems or DSL (Digital Subscriber Line). Once the bandwidth is available, video conferencing will become widely used in the home, especially for document conferencing when two employees must jointly work on one shared document. Of course, once video conferencing in the home becomes common, people will have to wear something other than their robe and slippers when they go to "work".

Judicial Applications

- The judicial system has found video conferencing to be a cost effective and productive technology for its needs. A number of counties have begun to install videoconference systems in jails and courthouses. These systems are used most commonly for "video arrangements", where a prisoner will go to a video conferencing room in the jail. Another system in the court room has dedicated cameras to show the judge, prosecutor and defense attorney. The prisoner can see all of the members in the court, and the information is simultaneously recorded with a split screen 4 ways, so a complete record of the proceeding is made. Tins provides several benefits to the court system, as it reduces the number of defendants who need to be transported from jails to court houses, reduces the number of defendants who need to be transported from jails to court house, reduces the over crowding of court house holding facilities, reduces the security risks associated with transporting and handling defendants, saves time and saves money. Most of these types of systems currently run over private networks using. Proprietary technologies, but they could be designed using H.323 technology over a secure intranet as well.

Campus Surveillance and Security

- To imagine scenarios where H.323 conferencing could be adapted as a surveillance/security tool, you really only need to look at situations where video is already being deployed for these purposes. Consider in particular those applications where connecting the video device to a LAN could enhance functionality and flexibility, or reduce costs. For instance, many campus and corporate environments have extensive LAN/WAN infrastructures that include both Internet and intranet connectivity. An H.323 surveillance solution would allow monitoring devices to be positioned anywhere there is a LAN connection and allow video from these sources to be viewed, recorded and easily integrated with other organizational data systems. This type of "surveillance" can extend beyond support for security applications. H.323 video conferencing terminals can also be placed in selected areas to provide "remote attendants", e.g., a "video receptionist", a "video test proctor", or "video lab support". The other end of the terminal would then reside on the desktop of the actual person performing the function, allowing them to monitor what is going on in the room and also "communicate readily when assistance or mediation is necessary. However, the ability to incorporate less expensive, more flexible systems on behalf of security and surveillance is not far away and, in many eases, may already be available.

Scientific Research

- If an engineer is studying aircraft wing design with several colleagues who are distributed around the country. He has implemented a large scale application on a parallel computing system at one of his sites (actually, it could be anywhere on the network). The person at that site can begin the application and click on data collaboration so that each of you can interact with the model as it runs and see the results as they happen. He is also using CAD software (which runs in an X-Windowed environment) to analyze the output further. Anyone can will start up the CAD software and click on application sharing. All of them can then view the structures and discuss what happened, what to try next, etc.
- Laboratory interactions involving experimental facilities as well as one or more people: A "virtual, laboratory" environment would be created by configuring all laboratory instrumentation to be compatible with automated control, and by augmenting network video conferencing capabilities with currently available, network windows based laboratory instrument control software.
- Video conferencing sought to find applications of it to **entertain** also. It can also be used for uniting family members at the time of separation. As on ongoing community service. San Juan College's Officer of Technology Services, in cooperation with Freedom Calls Foundation, utilizes this state-of-the-art technology to keep us troops in Iraq in touch with their families in the Four Corners.

Limitations

- Preparation time involved in the project is high.
- **Equipment compatibility**: Equipment used in video conferencing cannot be carried to all the places
- Locating video conferencing sessions on specific subjects can be a challenge.

Future uses

- What the future holds for this technology is as uncertain as any innovation it may be refined and honed to a better stale, or junked when a newer, better technique is developed.
- Already, though, the methodologies for further advances in this field are on the brink of readiness: surgeons being able to perform operations from around the world via video cameras, robotic manipulators, and remote controls; faster and more detailed procedures to present conferencing of a better quality;
- Mass produced imaging wares that would allow any person with a computer to be able to communicate with anyone else over a video line, even the video phone so long held clear by science fiction writers.
- Whether video conferencing remains for the duration or is put aside for something better, it currently holds the position of technology of choice.

TELE CONFERENCING APPLICATIONS

Teleconferencing

Two-way electronic communications that take place between two or more groups, three or more people who are nil in separate locations arc known more commonly as teleconferences. Teleconferencing accounts for another technological advancement that lows for a multiple person phone call to replace face-to-face meetings, letter exchanges, is/or multiple phone calls. It allows for increased efficiency and productivity in the business forum because it saves time and money, allowing dispersed members of an organization to hold meetings from any number of points around the world at the click of a button. Telecommunications links use both regular phone lines and satellite phone lines to connect users from almost any point on earth.

Video teleconferencing

The ability of businesses to be able to readily communicate with the rest of world (Both inside and outside its industry) plays an integral role in the company's survival and success. The technological advancement incorporates full color/full motion video feeds with high quality audio transmitted live, in real time, from one location anywhere in the world. This technological advancement is referred to as video teleconferencing.

Benefits to Video conferencing

Time saving equals more time to be productive

Business able to save big off travel expenses

Shorter meetings

Easier scheduling as a result of shorter meetings

Increased connection to the outside world

Software for Teleconferencing

Internet standards for tele-conferencing are supported by several software

T-120 Multipoint / Multicast Standard

H.324 Videophone Standard

H.323 Videophone Recommendation

ICQ - A program for Personal Communication Center

Open Buddy List Proposal

CU-SeeMe

CU-SeeMe is video chat software for interacting with others using full colour video, audio, types, text, and white board communications. With Cu-SeeME, you can make new friends, expand your professional antisocial networks, find people with similar interests, watch entertaining cybercast events and much more. There is a on-line directly to locate other CU-SeeMe users and a list of public reflector sites for live group interaction.

Control in Teleconferencing

- Dial out to one or more participants and join them to the call.
- Place participants in listen only more so you are the only one who can speak.
- Mute of un-mute your own line. If you are not speaking you may wish to use this feature to eliminate any interface or background noise.

- Lock you conference so no one else can join. This ensures confidentiality of your calls.
- End the conference call by disconnecting all lines.
- Get a count of how many participants is one the call.
- Hear a recording of the participants name as they enter the conference

Participants have two touch - tone commands available to them

Mute and un-mute their own line. This prevents background noise from disturbing the conference.

Reach an operator for help.

Uses of teleconferencing

- Some of the many uses for teleconferencing include
- Board and committee meetings
- Staff meetings
- Discussions with headquarters and chapters
- Dispersed annual meetings
- Staff Training
- Project reviews
- Continuing Education
- Member updates
- Press Conferences
- Special interest Group meetings
- New Service announcements
- Regional meetings
- Public information programme
- Trade fairs
- Public Education programmes
- Staff Development
- Reduce costly travels and travel time
- Hold larger meetings more easily
- Reach much wider audiences

- Decreases meeting time.

AUDIO CONFERENCING

The simplest form of teleconferencing is the use of amplified speaker phones to in groups of people at each end of a telephone conference. High quality audio reception and transmission can be achieved here.

Features of Audio conferencing AP Reservations

Provides the ability to schedule and conduct conference calls on the spot with no advance notice.

Conference now Calls

H Provides the flexibility to conduct an audio conference call, with up to 75 individual, anytime without a reservation. Permanent toll or toll free phone number and pass code will be provided.

Participant Notifications

Upon request, we can fax notifications to all your participations to notify them of upcoming meeting.

Broadcast Fax

Meeting agendas, preservations, or other materials can be faxed to all participants be the conference center prior to the cell.

Host Preference Database

As part of your Audio Conferencing Service, the Co-ordinator will create a Host file in our database that will track your billing information and user preferences by individual hosts.

RSVP line

Allows participants to call into an automated system and register for the conference call ahead of time. The information in the automated system can be customized specifically to meet your needs.

During Conference Features

Sub-conferencing

This enables dedicated groups to opt out of the main conference call and regroup as needed.

Broadcast / Listen only

Conservation can be limited to one or more dedicated speakers, others are in listen only mode.

Electronic Polling

The host can use this feature to track audience response, survey questions, or key points of a meeting. Participants simply key in their response on a touch tone and the results can be printed for the host.

Electronic Question and Answer

Participants who have questions are placed in a queue and called on one at a time Conference Coordinator. This feature is often used with Broadcast/Listen only.

Translation / Real Time Interpretation

Foreign language translator can participate in the conference call.

Post Conference Features

Transcription

Transcription of meeting is available to the host, either printed or on disk.

Conference Recording

A DTP, WAV file, MP3, CD, or cassette tape of the conference is available.

Post -Conference Participant List

A list of all conference attendees can be sent to the host of a conference call.

Digital Tape Replay

Allows users to dial in and listen to a taped replay of the conference call 24 hours day. Digital recordings can also be used without a conference call for recording important announcement that can then be accessed via a toll-free dial in number.

Translation

Post-conference written translation in over 60 languages and dialects is available.

Rich Audio Conferencing Features

Audio conferencing software includes powerful scheduling, notification, and integration components that make it superior to oilier conferencing software on the market. As e.g. result, its easier and more intuitive to learn and use so companies can maximize the benefits of its audio conferencing features enterprise wide.

Information Communication and Technology (ICT)

Information Communication and Technology (ICT) in simple term means, any product or system that communicates, stores and or processes information. Digital convergence allows the same channel to be used for data, voice and images there by enhancing potential for human interactions. ICT is playing a multi faceted role for changing the model of business transactions. ICT is facilitating new forms of interactions. Internet reduces the transaction cost and increases the market access. These create new windows of opportunity. ICT has an enormous impact on shaping the mindsets and attitudes of the society where the ICT applications will be applied to the traditional needs of the society.

Trends of development in ICT

Rural women form a very significant part of the productive work force in the agriculture sector in India and most of the developing countries. Women are central to the selection, cultivation, preparation and harvest of food crops. Women farmers also possess a unique knowledge of dairying, animal husbandry and fish farming and handle most of the work associated with it. Verma (1992) indicated that women are responsible for 70% of actual farm work and constitute upto 60 per cent of farming population. He also stated that role of women in agriculture is substantial and significant. In fact, historians believe that it was women who first started cultivation of crops and initiated the art of science and farming.

"Farm women continue to be the important stakeholders in farming activity of India. Presently, they constitute one third of the agricultural labour force and about 48 per cent of self employed farmers" stated Praveena *et al.* (2005). Furthermore, management and involvement of Indian women in farming enterprise has been on rise in recent years especially in better endowed rural region.

About 80 per cent of all economically active women in the country are engaged in the agricultural sector. Also, about 48 per cent of India's self-employed farmers are women. There are 75 million women engaged in animal husbandry as compared to 1.5 million men.

Contributions of farm women in agriculture cannot be ignored. The findings of Chaudhary and Sangram Singh (2003) support the above statement. They reported that women played considerable role in various agricultural activities such as pre-sowing, sowing and transplanting operations. Women's role was also reported in the preparation of compost and rising of vegetable nursery. Women's role was found to be maximum in hoeing and weeding practices on intercultural operations. Harvesting and post harvesting operations were reportedly more suited to the physical condition of women. Reaping of crops, storage of food grains, storage of seed and processing of grains were reported to be mostly done by women.

In India, women have a major place in agrarian scenario, Randhawa & Chandra (1993) stated that they play a significant and crucial role in various agricultural and allied activities besides fulfilling their responsibilities as home makers. Sangwan *et al.* (1996) reported that there is a distinct sphere of participation between males and females in farm and dairy sectors, the men were the planners and women implemented the activity. Their contribution is significant in the field of agriculture and allied activities as they are considered as backbone of food security.

It is an acceptable fact that women possess talent which could be harnessed for the productive purpose. But it is most unfortunate that the role of women in agriculture has been neglected. A holistic development of women requires empowerment in all aspects involving them as managers, planners, scientists, and technical advisories. This is supported by Radha Singh (2005) as "A large part of women's contribution in agriculture is often not recognized and is also unpaid".

Rural women from a very significant part of a productive work force in the agriculture sector in India and most of developing countries. Even then, they are negligible among the beneficiaries of knowledge, capacity and trained human resource to handle latest agricultural technologies. This is due to lack of access to technical agricultural information and education.

Women and Knowledge

The world is in the midst of knowledge revolution, complemented by opening up entirely new vistas in communication technologies. Recent developments in the fields of Information and Communication Technologies (ICTs) are indeed revolutionary in nature. Hundreds of millions of dollars are being spent on ICTs, reflecting a powerful global belief in the transformatory nature these technologies. ICT, when used as a broad tool for educating the rural communities especially women, heralds the formation of a new class of society – the knowledge society. Knowledge thereby becomes the fundamental resource for all economic and developmental actitivities.

The one resource that liberates people from poverty and empowers them is knowledge. Possessing knowledge is empowering, while the lack of knowledge is debilitating. The World Bank organized a forum called, "Voices of Poor" which got feedback from 60,000 people in 60 countries and concluded that people wanted access to knowledge and opportunities instead of charity to fight conditions leading to poverty", stated World Bank (2000).

Historically, the isolation of women from the mainstream economy and their lack of access to information because of societal, cultural and market constraints have led them to become distant from the global pool of information and knowledge. This distance is reflected in the levels of empowerment and equality of women in comparison to men and has enormously contributed to the slow pace of development of an agrarian country like India.

Gender based Technology dissemination in Agriculture

Agriculture is a world-wide critical strategic resource expected to double its production in 30 years to feed the growing needs of the population. Even now it continues to be the occupation and way of life for most of the Indian population. It is the lifeline of Indian economy. The sustainability of agricultural growth has been the focuses issue of the nation over the years.

Transfer of technology generally aims at speedy diffusion of novel agricultural information from the place of invention (lab) to the place of adoption (land). For about five decades the information and communication tool has been conventional and only Home visit and Personal approach has been adopted by the extension personnel to disseminate the technical information to farmers. This method has been found time consuming and failed to disseminate the right technology to farmers at the right time who are spread across the country like India with diverse needs and interests.

So far all developmental programmes right from the Community Development programmes to the recently launched Bharat Nirman were found to be men focused including the agricultural technology dissemination progrmmes. Despite multiple roles of women in agricultural operations and household chores, their work is generally underestimated and undervalued. This could be supported with the statement of Suman Singh and Puja Mathur (2005) as, "the status of women in general is much lower than that of their male counter parts, largely because of customary male dominance in society, inherent shyness of farm women and lack of opportunities for education and training.

Group Approach Vs Information and Communication Technology Tools

Information Communication and Technology (ICT) in simple term means, any product or system that communicates, stores and or processes information. Digital convergence allows the same channel to be used for data, voice and images there by enhancing potential for human interactions. ICT is playing a multi-faceted role for changing the model of business transactions. ICT is facilitating new forms of interactions. Internet reduces the transaction cost and increases the market access. These create new windows of opportunity. ICT has an enormous impact on shaping the mindsets and attitudes of the society for adopting the change process. The wider use of ICT will result in new social situations where the ICT applications will be applied to the traditional needs of the society.

Trends of development in ICT indicate that it follows the usual paths towards inequality and exclusion, putting women and girls at a disadvantage with regard to securing better livelihoods and other rights. Information technology is making a significant contribution in the empowerment of women. The women SHGs and their members are using ICT approaches in both farm and off farm activities. The government institutions are also engaged in the development and popularization of gender friendly tools and equipments. With this key idea in mind, development of user friendly multimedia for enabling farm women has been initiated as micro level action cum experimental research.

Interactive Multimedia Compact Disc (IMCD) – Multidimensional Tool

Any communication should be very effective in conveying the message such that it is properly understood and location specific, interactive, self-dependent and fast in delivery what is intended to be delivered. The traditional methods of communication lack one or the other above features and hence could not provide holistic information.

IMCD is an innovative self learning module highly useful to both literates and illiterates with greater control developed in local language. It offers an unprecedented potential of providing knowledge to local women communities on latest agricultural technologies. It could also be used to educate and empower women in all aspects of life like health, education, agriculture, home science etc.

It is multidimensional nature which could be used in areas of academic education, distance learning, self learning and transfer of technology tool. It is cheaper and affordable to rural women to gain knowledge in various aspects of life. It could be retained and viewed repeatedly as per the demand of the individual. It could also be an effective tool to bridge the digital divide among rural women prevailing in India.

Computer multimedia offer learners more complete and individual control over their learning. The main reason behind recommending the use of computer based multimedia system for farmers to facilitate interactivity and better understanding between individual learners and the subject matter. Multimedia tools are ideally suited to demonstrate complex and dynamic process that cannot be explained easily with conventional media and methods.

Features of Multimedia

Multimedia presentations may be viewed in person on stage, projected, transmitted or played locally with a media player. A broadcast may be a live or recorded multimedia presentation. Broadcasts and recordings can be either analog or digital electronic media technology. Digital online multimedia may be downloaded or streamed. Streaming multimedia may be live or on-demand. Enhanced levels of interactivity are made possible by combining multiple forms or media content.

Multimedia is a modern electronic gadget with a combination of text, graphics, audio and video allowing the end user to control the delivered elements. Aeron (1998) indicated that, if you are stimulated with audio, you will gain about 20 per cent retention rate that of audio visual is up to 30 per cent and in an interactive multi media presentation, where you are really involved, the retention is as high as 60 per cent.

Swanson *et al.* (1990) reported that person-to-person communication has traditionally ben the most important form of information transfer. Print media as well as radio and television were of supplementary nature because they frequently lacked target group or location specificity and information was not up-to-date. Revolutionary changes in communication technology have dramatically increased the speed and quality of information transfer and changed the role of extension workers in industrialized countries. Electronic communication systems may in part replace personal visits, and one of the major tasks of any agent will be of link her or his clients with other suppliers of information. The speedy diffusion of novel agricultural technologies could only be achieved through active involvement and participation of farm women with the help of IMCD. IMCD goes a step further in providing information to the individuals/farm women round the clock in local language using audio video clipping.

IMCD is increasingly attracting the attention of farmers including farm women, extension functionaries and educationalists as a way of much-needed speedy TOT in agriculture. It is fast becoming most successful tool for TOT and imparting education in agriculture. In fact, it is a motivational cum inspirational tool playing a significant role in sweeping changes in socio-economic development in the era of globalization.

In Tamil Nadu, the technology delivery system is managed by State Department of Agriculture. There are number of bottlenecks both in technology dissemination and user systems. Apart from this delivery system, the university scientists, KVK extension workers, development department officials and non-governmental organization workers are providing adequate information related to the latest know-how to farmers for their betterment. They are using print and electronic media such as television, radio, news papers, bulletins, folder etc., besides organizing method demonstrations and campaigns. Even though these media have been doing lot of works; the accessibility is again is the stand still. A few of them promisingly going forward only could the reap fruits of it. The larger sects, which include small and marginal women farmers, are the real underprivileged, to whom the information technologies role and help is needed in a big way.

Some important case studies are:

Jothi

1. My name is N. Jothi. I am 50 years old. My husband's name is Nagarajan. I am living in Mettuneerathan village of Vadipatti block. I have studied upto Seventh standard. I am a native of Solavanthan block and have been married to Mettuneeratahn. Ours is a joint family with six members. The main source of income to our family is agriculture and salary of my son, working as a lecturer in private Engineering college in Viruthunagar district. Myself and my husband are actively involved in agricultural activities. We own 9.5 acres of Vaigai cananl irrigated land and cultivating Paddy. I am one among the members of TANWA (Tamilnadu Women in Agriculture) group organized by Department of Agriculture and a member in SACRED Non Governmental Organization (NGO). I have gained good exposure on banking activities due to my engaging in NGO activities.

We are suffering with yield losses in paddy due to rat damages. We are unable to manage the rat damages. It was at this stage research scholars from Departmetn of Agricultural Extension and Rural Sociology, Agricultural College and Research Institute, Madurai, visited our villages and conducted various Participatory Rural Appraisal (PRA) exercises such as village resource mapping, seasonal calendar and pair-wise matrix ranking to identify various technological needs felt by the paddy growers of wet land eco-system. At the end of PRA exercise, we could come out with valuable technologies such as System of Rice Intensification, Integrated Plant Protection measures in paddy, Eriophyid mite control in coconut, Hybrid seed production technology for Paddy, Mushroom production technologies, Vermi compost production technology and Drip based sugar cane technologies.

Being a paddy grower for several years, I felt that plant protection measures on paddy Eriophyid mite control and rate management and seed production technologies in paddy are much needed technologies. I also felt that vermin compost production techniques is the need of the hour as it offer the necessary amount of organic manure and small business venture.

After a month of need assessment, research scholars gave hands on exposure training to the farm women on the developed IMCD modules.

A hand on exposure has been organized in our village nearly fifteen days. IMCDs developed on various felt needs have been exposed to us. Before exposure, I have not been aware of computer aids and its applications in agriculture. But the present exposure to IMCDs on various agricultural technologies has motivated me to participate in the computer aided learning. After attending this training I feel that I have gained considerable awareness and knowledge on agriculture technology.

At the initial stage, I found it very difficult in operating the IMCD, but after a period of fifteen days I became confident in using the tool. Realization of the importance of agriculture through IMCD, I enthusiastically involved myself in various activities such as data collection, pre-testing and post-testing of knowledge test questionnaire on IMCD modules etc. The various qualitative data collection activities viz., expenditure analysis, food calendar and savings preference ranking etc., too came under my keen participation.

The text material, photographs, audio and video clippings has helped me a lot to understand the technology without much hardship. This orientation programme has impressed me much and the way in which the organizers clarified all our doubts has been admirable. I have also been impressed by the plant protection measures on Brinjal which would enable me in avoiding a worse situation like that of selling vegetables from my field as animal feed last year. Thus, the IMCDs on new technologies, if provided, would help farmers like me in reducing wastage of resources as well as in increasing the yield.

The interactive features such as audio, video and text facilities used in the process of orientation have been satisfactory. While it has been difficult to visualize the symptoms of plants, disease and nutritional disorders the IMCDs have made it easy with the use of various photographs.

Besides other sources of information, IMCD offers authentic as well as wholesome information. Giving computer learning to the needy farmers and rural farm women would empower them to access information. At the same time continuous practice has to be given in computer application so as to enable the farmers to retain the gained knowledge and adopt it.

Now-a-days women too are competent enough to gain knowledge through computers and improve their way of living. Their keenness in welcoming a new technology an to empower them by various progressive farmers would definitely motivate those who are new to this field.

2. Case study of Mrs. Pothammal

My name is Pothammal, residing in Mettuneerathan village of Vadipatti Block, Madurai district, Tamil Nadu, aged 55, I have two sons who were married a daughter. She owns one and a half acre wetland. I live in a thatched house with mud floor not with much infrastructure facilities. I was holding membership in many farm women organizations and clubs for a period of twenty years and still acting as a leader of Tamil Nadu Women in Agriculture (TANWA) group.

It is an active and enthusiastic farm woman having more experience in farming for the past 30 years. It is the contact farm women for many officers like state department of agriculture, Krishi Vigyan Kendras and from other women clubs. I am the one who initiated the exercise with much involvement and acted as a role model for the others to follow her. I also narrated the whole exercise by listing out the various resources available in the village. Furthermore, I extended full cooperation till the end of the project.

Not many came forward to know about IMCD and its operation during the orientation training. But I was the only women who convinced all the other rural farm women of her village making them recognize the importance of IMCD and its role in TOT. The one valid reason which I told to convince farm women was, she used to travel miles together to get to know about new farm technologies. For e.g., Agricultural College and Research Institute, Madurai or to other districts of Tamil Nadu. Moreover, I also told that they would be able to learn what is computer and its usage. But now, I told to her group members and others that a golden opportunity was given to them to learn about computer and farm technologies at their door steps.

Before giving orientation, I knew nothing about computer, But after the orientation training, she was able to operate computer hassle-free. I expressed that she never expected a good chance like the intervention of the project to teach computer and its usage at this age. I felt much happy about it and thanked the research team and the agricultural college of Madurai district for having taken up enormous efforts implement the project.

I gained knowledge on latest farm technologies on Mushroom Technologies, Vermi compost Production, Integrated Plant Protection Measures in Rice, System of Rice Intensification, immediately after exposing the developed IMCDs. She opined that IMCD is a very useful tool for transfer of technology as it conveys message in local language (Tamil) with audio, photo and video. She expressed that she is not that much literate but was able to learn a farm technology through IMCD in a more comprehensive and easy way. I also told that she felt ease is operating the IMCD as it gives self instructions on how to navigate the whole IMCD. In fact, she told that IMCD is going to revolutionize the agricultural extension in farm technology dissemination.

Utility of IMCD

- The interactivity, incorporation of video and sound, the ease in operation are the traits of Multimedia which will be very much useful for even illiterate and semiliterate farm women to learn practices.
- Any farmer who wants to have hands on information about specific doubts over "Interactive Multimedia Compact Disc on Identified Technologies" could easily make use of Multimedia without any extension person's facilitation at any time and anywhere cutting across the spatial and temporal hurdles.
- The ICT revolution and its link with dissemination of agricultural technologies has been a hot topic which is widely debated upon. The extension personnel would find Multimedia as a timely, cost effective, novel and efficient tool for dissemination of GAPs related to other agriculture and allied activities.
- The results of this project revealed that the effectiveness of Multimedia as a dissemination tool. This can be recommended as one of the successful mode of transfer of any agricultural technology in general and complex technology in particular.
- The project employed a group based approach by creating Farm Women User Groups. This approach has been time tested and even after the project is completed the groups will sustain which is another tangible impact the project has created.

Agricultural Extension and ICT

Agricultural Extension in the current scenario of rapidly changing world has been recognized as an essential mechanism for delivering information and advice as an input into modern farming. In spite of the huge potential of the Information Technology that can be harnessed for agricultural development, only a few isolated projects have been initiated in India. Interestingly, many of these projects were initiated by non-governmental organizations (NGOs), private organizations, co-operative bodies and governmental organizations other than agricultural departments. This indicates the non-readiness of agricultural departments to use ICTs in their programmes.

Most of the ICT initiatives in India are location specific, isolated and confined to small area, providing services for a specific aspect of agricultural development. The efforts have to be institutionalized in order to draw lessons and develop insights into these differential approaches of the projects. The project is an important prerequisite for developing a social strategy.

The effective use of ICT's requires capacity building. The ICT's play a crucial role in operationalization and spreading extension reforms. In developing any system of Information Technology for transfer of agricultural technology, the farmer is to be kept in the centre stage. He is not to be treated as a passive recipient but rather as a player, generator and user of knowledge system. Information and Communication Technology would play a supplementary role in Agricultural Extension for financial sustainability.

III TRANSFER OF TECHNOLOGY

Chapter – 11

Problem Tree Analysis

In most countries, extension systems have now been operating for some time and problems within them will often require a refinement of the existing system rather than design of a new system from scratch.

The purpose of this chapter is to provide agricultural extension managers with a simple approach to analysing common extension problems. In most cases, suggestions are then given for possible actions to consider in investigating the causes of the problem further and ultimately in overcoming it. However the primary purpose of the chapter is not 'to give solutions to all known problems' but to assist the extension manager in analysing problems to find their root cause (or causes). The chapter is designed to be simple and easy to use, and thus cannot do justice to the complexity of all the problems and situations faced by extension personnel.

Many different approaches might be used to organise this process and information. The chapter has tried in all cases to begin with noticeable problems and by analysing these through an hierarchy or problem tree, help users to analyse and reach their own conclusions as to the best solution.

It is important to recognise that all agricultural extension systems necessarily differ to some extent due to climatic, cultural and economic differences between countries and regions. Within a given system the problems and their solutions are to some extent unique to the particular people and circumstances under consideration.

This chapter should therefore not be seen as a recipe book - it cannot provide specific answers to problems. Rather it aims to provide a tool to the extension manager to help him or her correctly analyse the situation and pin point the reason or reasons behind problems and hence find possible solutions. What may be a sensible practical solution in one area could be entirely inappropriate in another. Suggestions for actions should in all cases therefore be seen as just that - suggestions. These ideas are intended to assist the extension manager in thinking of his or her own appropriate solutions according to circumstances. In most situations there will be a number of possible courses of action or further investigation which could be undertaken. Also in many cases, problems may have more than one cause and will not exist in isolation but have a number of other related problems. In all cases, having arrived at a possible cause for the problem, the user of the chapter should ask whether the problem could also relate to another area and consider alternative causes and actions to take.

While the chapter covers analysis of a large number of potential problems, in all cases the user is urged to consider whether there may be other reasons for the problems not listed and not covered elsewhere in this chapter, i.e. at each branch of the problem tree, the user should consider whether the issue is covered or if a further branch is needed in their particular case. In most cases, problems should be further analysed by the user beyond the points given.

The chapter is divided into three sections reflecting the various actors in an extension system as seen from the point of view of the middle level manager:

- The rural people
- The field extension agents
- The extension manager him/herself.

Within each of these areas, a number of common problem areas are identified. These problems are then further subdivided in an hierarchical way leading to possible root causes from where suggestions are made for possible actions to take.

In all cases, the chapter begins with the assumption that the agricultural extension system exists to achieve certain objectives and therefore the problem areas are those which prevent the system from achieving its objectives. While it may be the case that the objectives of the extension service itself are inadequate or impractical, this is outside the scope of this chapter. Such issues deal with the design of the extension system itself which is covered elsewhere.

If all problems therefore can be seen as stemming from lack of progress in achieving extension objectives, it follows that many of the issues commonly thought of as problems are not necessarily a problem unless there is no alternative way to achieve the objective except by overcoming that obstacle. For example, lack of transport is not in itself a problem except in as far as it prevents the extension system from reaching the farmers with who they work. Inability to reach the farmers with the advice and assistance needed is the fundamental problem and this could be overcome in a number of ways such as for example better use of mass media, group extension approaches, public transport, setting up extension offices closer to where they are needed or making better use of local people as extension agents themselves.

For any given problem:

1. Identify which of the three groups is initially considered to be where the problem lies i.e.
   - Rural People (farmers)
   - Extension workers
   - Extension management.
   - (further investigation may show that the problem is actually with one or both of the other groups)
2. From the Table of contents, identify which problem area or areas most closely matches the problem under consideration.
3. Turn to the appropriate section of the chapter and follow through the hierarchy to trace possible causes.
4. Some levels of each hierarchy are then listed as specific problems and possible actions to take are proposed. Suggestions are intended as examples of alternative course of action only - in any given situation it is likely that there is more than one possible course of action.
5. The map pages following give an overview of the hierarchy of problems which may also be helpful in identifying possible causes of problems and suggestions for dealing with them.
6. There is inevitably some overlap in both problems and possible causes. In all cases consider whether the problem may lie in other areas which need to be considered in analysing the problem which may suggest quite different causes and hence different actions to be taken. It is

ultimately up to the user to decide the best solution. Clear analysis of the problems is the first step in finding solutions which in most cases will need to be specific to the situation of concern.

RURAL PEOPLE (FARMERS) CENTRED PROBLEMS

Rural People centred problems are those which are initially perceived as directly related to the farmers with whom the extension service works. Further investigation may, however, show that the root of the problem is actually not with the farmers themselves, but with the field workers or with extension management or even with overall extension objectives.

This category includes all issues related with adoption, implementation of field activities and participation of farmers and villagers in extension activities.

A fundamental purpose of most extension services is to promote advice and innovations which will improve the production levels, storage or marketing of agricultural products. There can be many reasons for the low adoption of extension advice - some of which are listed in this section. These can range from inaccurate/inappropriate advice to simple lack of awareness of the idea - i.e. poor extension communications

A survey, or analysis of data collected through the extension service, may highlight that the level of adoption of extension advice is below the targets set. However, before considering any action, it may be necessary to review what the information on low adoption means. Has only adoption of a full package of recommendations been considered? If so, this may exclude possibly numerous cases of partial adoption. Any action to be taken should be very different if there is widespread partial adoption than if there is no adoption at all.

Group discussions are often useful and a starting point might be to call a meeting with the field workers in charge of the area where the low adoption has been recorded, and begin by discussing with them the problems they might have noted. It should be made clear that there could be many different reasons why adoption is low and that none of them are being blamed for the results. Stressing a team approach in working together to understand the situation better and improve the effectiveness of the work can be helpful. Discuss the problems they have identified and ask them to further investigate the issues while in the field.

Among issues to investigate further in order to understand the actual reasons of the low adoption rates are:

- do farmers know about the extension advice?
- is no-one in the rural community aware of the advice or are only some people unaware of it?
- are there any common factors among the people that are not informed of the technology (e.g. they are not part of the key contact person's family, they are all women, or all older farmers, or just people living in a certain area, or belonging to a certain ethnic group, etc.)
- is the awareness low for all the extension advice introduced, or does it relate only to a particular type of technology/message?

A major reason for the low level of adoption of recommendations can be that the farmers for whom the advice is intended are not being reached, and are not aware of the advice at all. Consider reviewing the channels of communication - are they appropriate to the intended group? What alternative means of communication might be used?

Possible reasons for low awareness of extension advice are that... communication channels used are inadequate or unsuitable for the intended audience. The choice of communication channels should consider the literacy level of the audience, the language or dialect spoken, the access to the media used (e.g. do the intended audience have access to radio, television, newspapers etc.). the channels are not being used effectively, e.g. the wrong time, location, duration is used to put across the information such as a radio broadcast aimed at villagers when they are away from the radio, the information is distorted. If information does not reach the intended farmers directly, it will inevitably be altered in some way by intermediaries. This happens even when the intention is to pass on the same message. The more directly a message can reach the intended people, the less the chance for distortion, the message is not seen as relevant - are illustrations pretested? Is the message appropriate from their point of view?

Problem: (low adoption)

extension worker spends too little time in the Held

Suggestions

If extension staff spend the majority of their time in the office, they cannot be reaching many of the farmers directly. There are many reasons why extension staff do not visit the rural people as often as they should-see also extension worker spends too much time in the office.

Key contact persons may receive information from the extension workers but not pass it on as expected. Among the reasons behind this there could be that:

- Key contact persons are not representative
- Their fields are too far from other villagers fields
- They do not welcome other villagers onto their fields
- They do not wish to pass on advice which may give other villagers an advantage
- They do not consider the advice to be useful to them - though it might be useful to others.

If it is found that other farmers are not aware of advice passed on to the key contact persons, the situation needs further investigation. Consider the other suggestions under

General low awareness of recommendations

Problem: (low adoption)

some key contact people don't pass on advice/information

Suggestions

- review use of those key contact people who do not pass on advice? emphasize contact with those that do pass on advice/information to others? Appoint additional key contact people if necessary?

It may be that some key contact persons need to be appointed due to their position in society. The emphasis may then need to be to ensure that these farmers do not monopolise too much of the extension service's time. If they are influential in the society, it may be that the extension service cannot operate effectively if they are not included as key contact people.

Problem: (low adoption)

Only some information passed on Suggestions

If only some information or advice is passed on, further investigation is needed to find out what information this is and what is the reason for other information not being passed on. Are there perhaps better ways to reach other farmers? Consider reviewing information which is not passed on. Is the advice given actually useful for the farmers? Is it technically, economically and socially feasible for the key contact persons? For the other farmers? Does the information give the contact persons an advantage over others that they would not wish to pass on? Consider also what other communication methods might be used to ensure a wider spread of information?

The next stage of the analysis is therefore, to identify the reasons behind this. It is important to involve farmers directly and find out from them what the reasons are, instead of assuming some issues might represent an obstacle to adoption. It may be useful to arrange a meeting with farmers to discuss what aspects of the extension package they find difficult to adopt. Is adoption limited by the lack of inputs? funds? time? skills? knowledge? labour? Or is the risk involved in the adoption too high? Are there cultural or social reasons limiting the adoption? Are there any other reasons acting as a constraint to adoption?

Inappropriate advice may be a advice which is technically inappropriate i.e. advice which for some reason is technically unsuitable for the particular area of concern. For example, certain tree species might be recommended but in fact be inappropriate for the local site. Advice may also be economically inappropriate i.e. unsuitable because the villagers concerned cannot afford the inputs required - or perhaps are not sufficiently convinced that the risk is worthwhile. Thirdly the advice may be socially inappropriate - that is, it goes against local ideas, beliefs or customs. It may also go against religious beliefs of the community concerned. Inappropriate recommendations are sometimes made by extension workers because not enough effort has been put into finding out what rural people's problems are and checking that proposed solutions are appropriate before promoting them.

In some cases, research organisations develop techniques and technical packages which are then expected to be promoted by extension services without first checking on their appropriateness with those whose livelihood is at stake i.e. the farmers who are expected to implement the recommendations with their investment in time and resources.

Discussing the issue of low adoption with extension workers and farmers might highlight that there is a satisfactory level of awareness of the advice. Farmers are aware of the advice but for some reason are not willing to take it up.

Problem: (low adoption)

Lack funds

Suggestions

- Check costs of proposed practice including opportunity costs-will anything have to be foregone to pay for the proposed practice?
- Check credit availability for rural people's group?
- Consider subsidized credit?
- Check risks involved?
- Could the advice be adopted gradually with less need for major funds at any one time?
- Could group ownership be included?

As with inputs, consider also whether the lack of funds are seasonal or continuous. What is the cause? Is erratic production or low yields the problem? What could be done to address these problems?

- part of the task of most extension systems is to increase the knowledge and understanding of rural people in the areas of concern. If advice given is not practical because the farmers
- lack the necessary knowledge or understanding, the extension service may be placing too little emphasis on this area. To help ensure that ideas are understood, extension staff needs to emphasize interaction during demonstr-ations and participation of villagers during all extension activities.

Depending on the extent of this problem, training courses may be needed for some extension staff to improve their skills in communications.

Extension worker lack of skills

Poor communication/demonstration skills

Problem: (low adoption)

Poor extension worker communication/demonstration skills

Suggestions

- other suggest consider whether extension workers need training on
- communication skills in general

- encourage farmers to take part in/practice more during demonstrations
- use more learning by doing exercises?
- listen more/speak less during group discussions?
- repeat ideas and demonstrations?
- ask those who do understand/do the recommended practice correctly to demonstrate/explain to others?
- encourage villagers to say so if they do not feel confident or do not understand/agree?
- See also B6 extension worker lack of skills

Problem: (low adoption)

Literacy problems

Suggestions

Low literacy rates can be a barrier to extension work since written leaflets and handouts are then of little or no use. However there are various ways in which the problems can be overcome. Drawings and illustrations if adequately pretested for understanding with the intended group can help communication with a low literate audience. If illustrations are not pretested, they can either have little or no effect on increased understanding - or at worst even give a wrong or conflicting message. People learn to read pictures much as they learn to read words and representations of size, perspective, angle and 'front and back' for example are not necessarily understood if they have not been seen before. It is also common for people unused to pictures to try to recognise a specific real person, animal or building in a drawing rather than just a symbolic person etc. It is essential that pretesting of messages be conducted if illustrations are to be used with little explanation.

Radio and cassette recordings, which are relatively cheap to spread messages consistently and raise discussions, can also prove useful. Slides and cassette tape are often practical ways of explaining situations - though are often limited in use by the lack of slide processing facilities available to the extension services. Video similarly can be effective in working with illiterate populations but in many cases is still too expensive and inaccessible.

Rural people's (farmers) problems not correctly identified

Investigations into the reasons for the limited appropriateness of extension advice may reveal that rural people's problems were not correctly identified. An extension service has little purpose if it does not respond to local problems, nevertheless it is common for research to be conducted with little appreciation of the intended end users of the results. Extension may then be blamed for not successfully promoting an inappropriate recommendation.

If it is found that local problems were not correctly identified, questions which would need to be answered before proceeding include:

- is the research team working on the wrong priorities?
- are farmers sufficiently involved in the identification of problems?
- are they willing to cooperate?
- do they have a misunderstanding of the reasons why they are asked about their problems e.g. they see extension workers as government representatives investigating their private affairs or finances.
- Or they see extension as providing them with gifts if they give the right answers - suggest that they have particular needs which may not be real, (such perceptions would lead them to distort the information they give).
- do adequate channels for accessing farmers' problems exist? in the day to day advisory activities do extension workers identify the problems correctly?

Problem: (low adoption) insufficient effort in identifying local people's felt needs

Suggestions: other hold open village meeting to discuss problems? Facilitate local brainstorms groups on problems? increase efforts to involve farmers in defining problems? use mass communications to increase awareness of farmers role in identifying problems? explain to farmers the reasons for wanting to know their problems?

Problem: (low adoption) local people not willing to share information with any extension workers an extension worker and the local people. Extension officers may also be seen as representing the government and if their role includes enforcement of regulations this can make relations with farmers difficult.

Suggestions

- ❖ hold group meeting to discuss problems?
- ❖ contact village leaders/key contact persons to analyse reasons?
- ❖ ensure the reasons of asking for information are clear to farmers?
- ❖ find examples from nearby areas, of successful extension packages developed after good problem identification?

Often the root cause of low adoption rates is the inappropriateness of advice, due to lack of attention to identifying the felt needs of farming community in the first place. Research or extension will frequently design programmes to promote a particular message without discussion with rural people or ensuring that it represents a priority need. Unless action is taken to closely link research and rural people, the research results may well be of little relevance to them and hence adoption rates will remain low. For local people to express their needs more forcefully, it is often useful to help organise groups at the local level.

Local people may not be providing the information needed because of the limited contact with extension workers. To build up trust and confidence takes time and frequent contact between extension worker spends too much time in the office

- ❖ other suggestions

Due to perceived lack of competence

If extension workers do not have the necessary skills or knowledge to carry out their work, they are unlikely to gain either the trust or confidence of the local people. In this case they may need further education and training or perhaps are not appropriate to that particular job or area.

It may however, also be the case that the extension worker is perfectly competent technically but is not good in communications skills and hence is perceived as incompetent by local people. Other reasons for local people not wishing to discuss problems with a particular extension worker may be due to their age, sex, ethnic group, education and background, religion or other social factors.

Problem: (low adoption) farmers not willing to discuss problems with particular extension worker due to perceived lack of competence

Suggestions

- other suggestions
- if due to lack of skills - may need to provide skills training
- if lack of knowledge, may need further education, or consider
- whether appropriate for the particular post. Is the job itself
- clear?
- if lack of experience, the problem may be temporary - does the situation improve with time?
- is the extension worker actually competent but not good in communication skills? Consider communications training

See also

Extension worker poor communication skills

Extension worker lack of skills/knowledge

Problem: (low adoption) farmers not willing to discuss problems with particular extension worker due to other reasons (not due to competency)

Suggestions

Other reasons for farmers not wishing to share information with, a particular extension worker may be due to the social other suggestions characteristics of the particular extension worker and the local people with whom he/she is expected to work.

Farmers may also misunderstand why they are being asked about their problems especially in extension services where they have en enforcement and policing role. In such cases farmers may be reluctant to provide information to extension officers and it may be useful to review whether such a role is compatible with other duties of the extension officers.

Unclear extension objectives

- if the problems are due to age, sex, ethnic group etc. of the extension worker, consider changing area of duties?
- if due to higher education level, social class etc., consider whether the extension worker may need training and practice in participative communications skills?

FLOW CHARTS

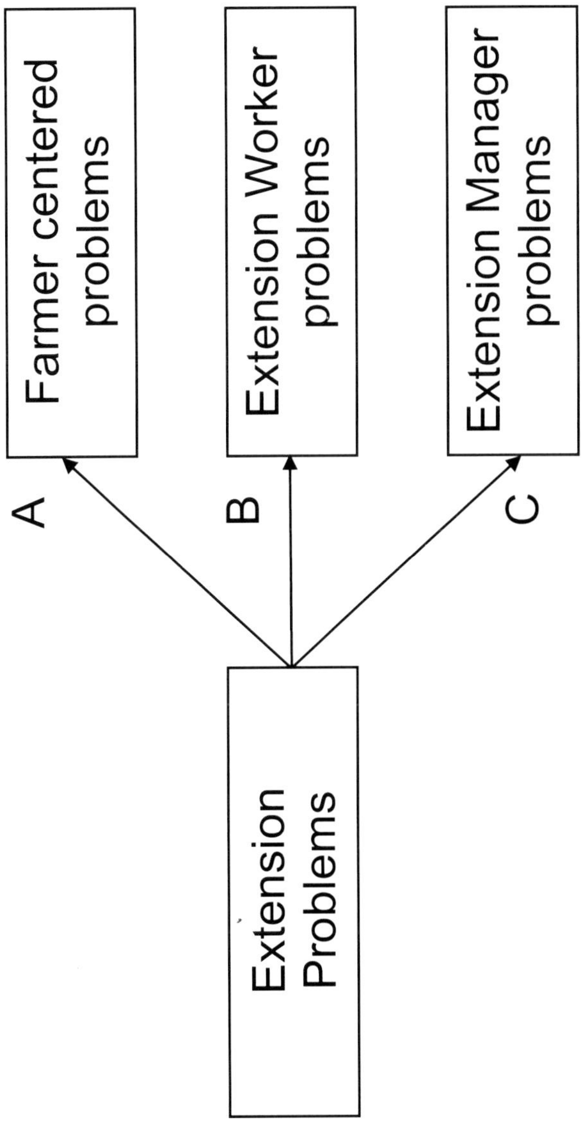
Extension Problems
A
B
C
Farmer centered problems
Extension Worker problems
Extension Manager problems

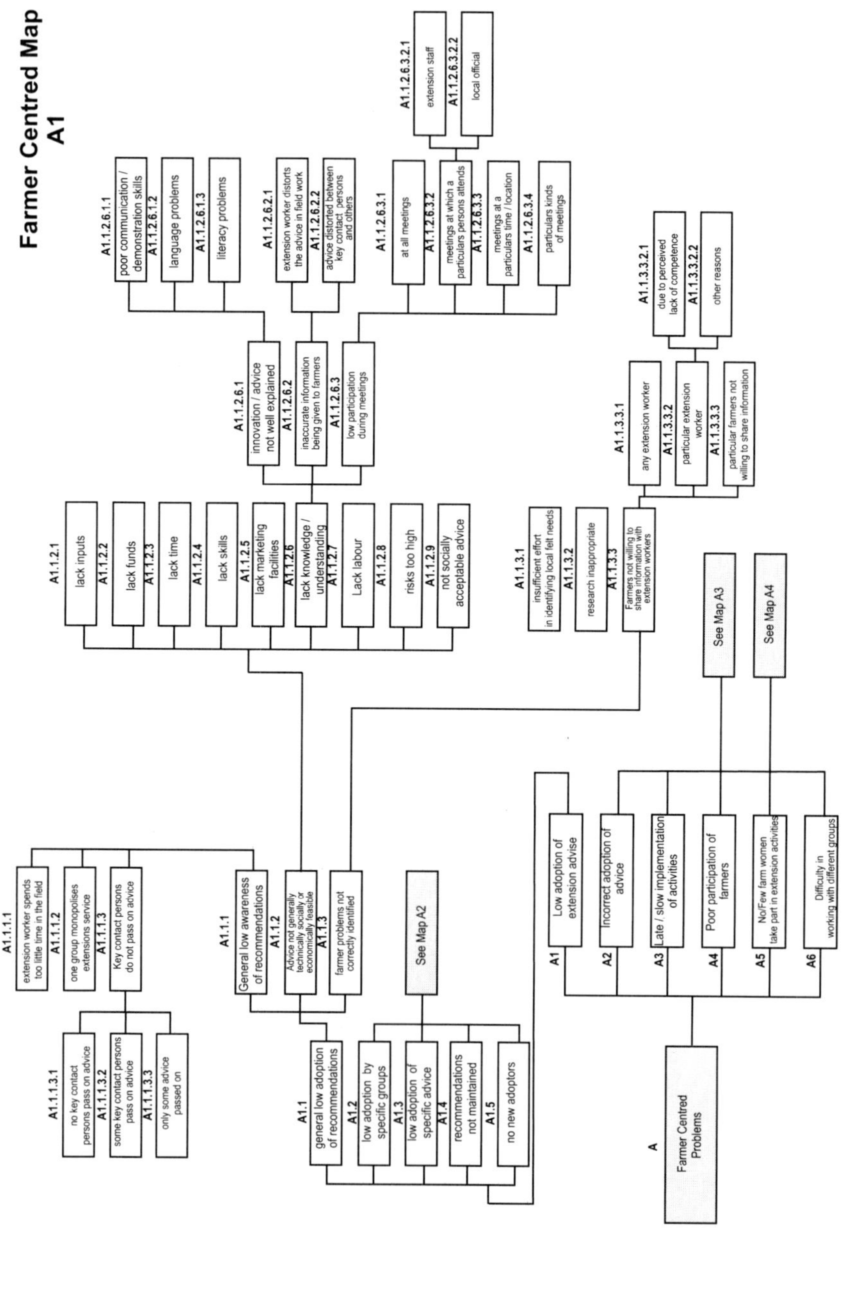
Farmer Centred Map
A1
A
Farmer Centred Problems
A1
Low adoption of extension advise
A2
Incorrect adoption of advice
A3
Late / slow implementation of activities
A4
Poor participation of farmers
A5
No/Few farm women take part in extension activities
A6
Difficulty in working with different groups
See Map A3
See Map A4
A1.1
general low adoption of recommendations
A1.2
low adoption by specific groups
A1.3
low adoption of specific advice
A1.4
recommendations not maintained
A1.5
no new adoptors
See Map A2
A1.1.1
General low awareness of recommendations
A1.1.2
Advice not generally technically socially or economically feasible
A1.1.3
farmer problems not correctly identified
A1.1.1.1
extension worker spends too little time in the field
A1.1.1.2
one group monopolises extensions service
A1.1.1.3
Key contact persons do not pass on advice
A1.1.1.3.1
no key contact persons pass on advice
A1.1.1.3.2
some key contact persons pass on advice
A1.1.1.3.3
only some advice passed on
A1.1.2.1
lack inputs
A1.1.2.2
lack funds
A1.1.2.3
lack time
A1.1.2.4
lack skills
A1.1.2.5
lack marketing facilities
A1.1.2.6
lack knowledge / understanding
A1.1.2.7
Lack labour
A1.1.2.8
risks too high
A1.1.2.9
not socially acceptable advice
A1.1.2.6.1
innovation / advice not well explained
A1.1.2.6.2
inaccurate information being given to farmers
A1.1.2.6.3
low participation during meetings
A1.1.2.6.1.1
poor communication / demonstration skills
A1.1.2.6.1.2
language problems
A1.1.2.6.1.3
literacy problems
A1.1.2.6.2.1
extension worker distorts the advice in field work
A1.1.2.6.2.2
advice distorted between key contact persons and others
A1.1.2.6.3.1
at all meetings
A1.1.2.6.3.2
meetings at which a particulars persons attends
A1.1.2.6.3.3
meetings at a particulars time / location
A1.1.2.6.3.4
particulars kinds of meetings
A1.1.2.6.3.2.1
extension staff
A1.1.2.6.3.2.2
local official
A1.1.3.1
insufficient effort in identifying local felt needs
A1.1.3.2
research inappropriate
A1.1.3.3
Farmers not willing to share information with extension workers
A1.1.3.3.1
any extension worker
A1.1.3.3.2
particular extension worker
A1.1.3.3.3
particular farmers not willing to share information
A1.1.3.3.2.1
due to perceived lack of competence
A1.1.3.3.2.2
other reasons

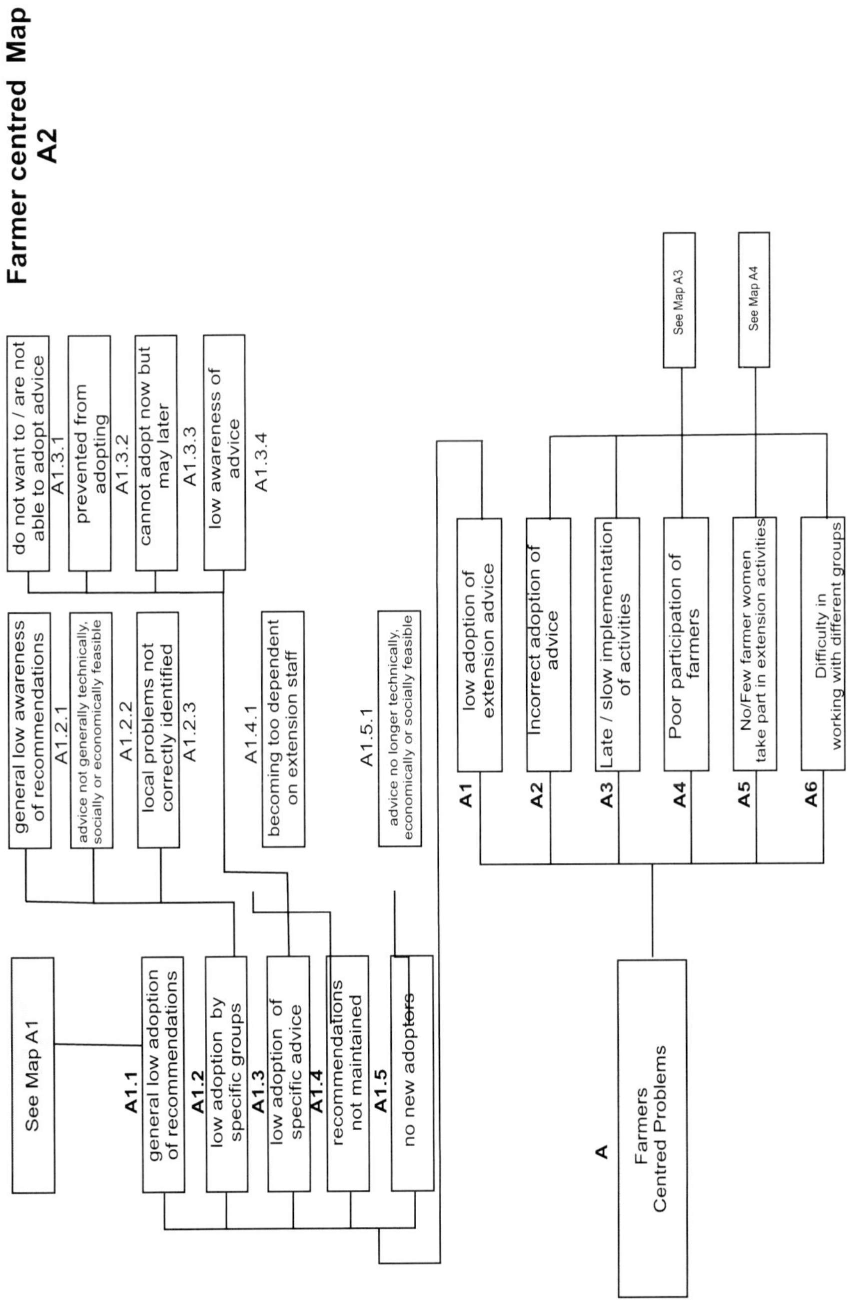
Farmer centred Map
A2
See Map A1
A1.1
general low adoption of recommendations
A1.2
low adoption by specific groups
A1.3
low adoption of specific advice
A1.4
recommendations not maintained
A1.5
no new adopters
general low awareness of recommendations
A1.2.1
advice not generally technically, socially or economically feasible
A1.2.2
local problems not correctly identified
A1.2.3
do not want to / are not able to adopt advice
A1.3.1
prevented from adopting
A1.3.2
cannot adopt now but may later
A1.3.3
low awareness of advice
A1.3.4
becoming too dependent on extension staff
A1.4.1
advice no longer technically, economically or socially feasible
A1.5.1
A
Farmers Centred Problems
A1
low adoption of extension advice
A2
Incorrect adoption of advice
A3
Late / slow implementation of activities
A4
Poor participation of farmers
A5
No/Few farmer women take part in extension activities
A6
Difficulty in working with different groups
See Map A3
See Map A4

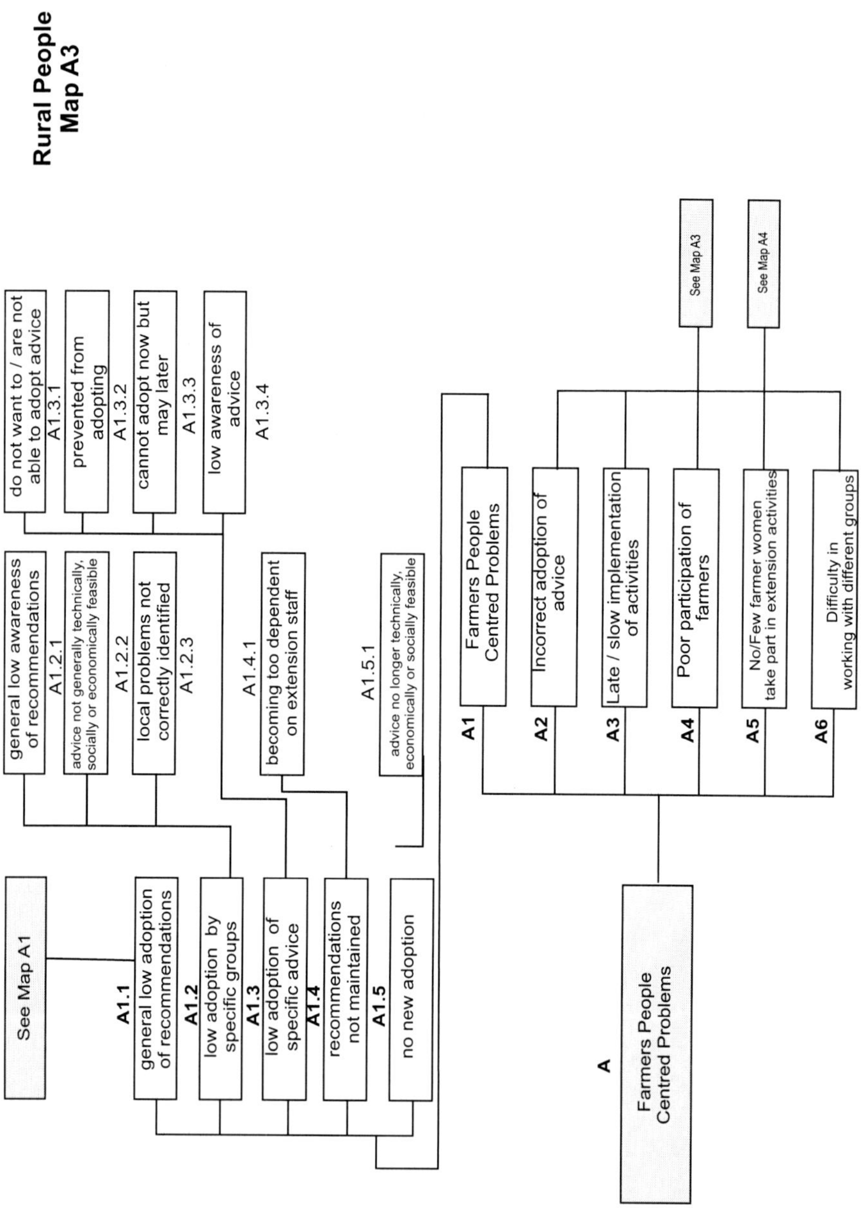
Rural People
Map A3
A
Farmers People Centred Problems
A1 Farmers People Centred Problems
A2 Incorrect adoption of advice
A3 Late / slow implementation of activities
A4 Poor participation of farmers
A5 No/Few farmer women take part in extension activities
A6 Difficulty in working with different groups
See Map A3
See Map A4
See Map A1
A1.1 general low adoption of recommendations
A1.2 low adoption by specific groups
A1.3 low adoption of specific advice
A1.4 recommendations not maintained
A1.5 no new adoption
general low awareness of recommendations
A1.2.1
advice not generally technically, socially or economically feasible
A1.2.2
local problems not correctly identified
A1.2.3
becoming too dependent on extension staff
A1.4.1
advice no longer technically, economically or socially feasible
A1.5.1
do not want to / are not able to adopt advice
A1.3.1
prevented from adopting
A1.3.2
cannot adopt now but may later
A1.3.3
low awareness of advice
A1.3.4

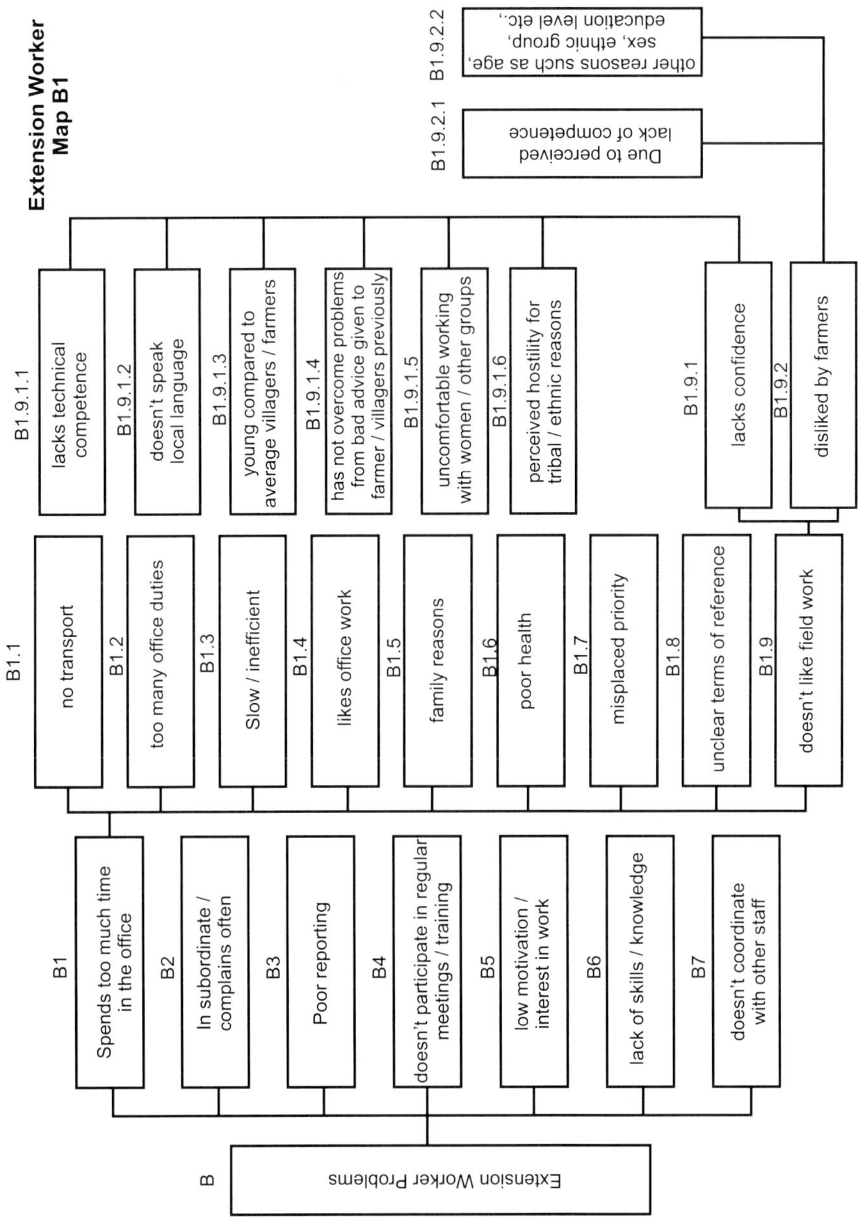
Extension Worker
Map B1
B
Extension Worker Problems
B1
Spends too much time in the office
B2
In subordinate / complains often
B3
Poor reporting
B4
doesn't participate in regular meetings / training
B5
low motivation / interest in work
B6
lack of skills / knowledge
B7
doesn't coordinate with other staff
B1.1
no transport
B1.2
too many office duties
B1.3
Slow / inefficient
B1.4
likes office work
B1.5
family reasons
B1.6
poor health
B1.7
misplaced priority
B1.8
unclear terms of reference
B1.9
doesn't like field work
B1.9.1
lacks confidence
B1.9.2
disliked by farmers
B1.9.1.1
lacks technical competence
B1.9.1.2
doesn't speak local language
B1.9.1.3
young compared to average villagers / farmers
B1.9.1.4
has not overcome problems from bad advice given to farmer / villagers previously
B1.9.1.5
uncomfortable working with women / other groups
B1.9.1.6
perceived hostility for tribal / ethnic reasons
B1.9.2.1
Due to perceived lack of competence
B1.9.2.2
other reasons such as age, sex, ethnic group, education level etc.,

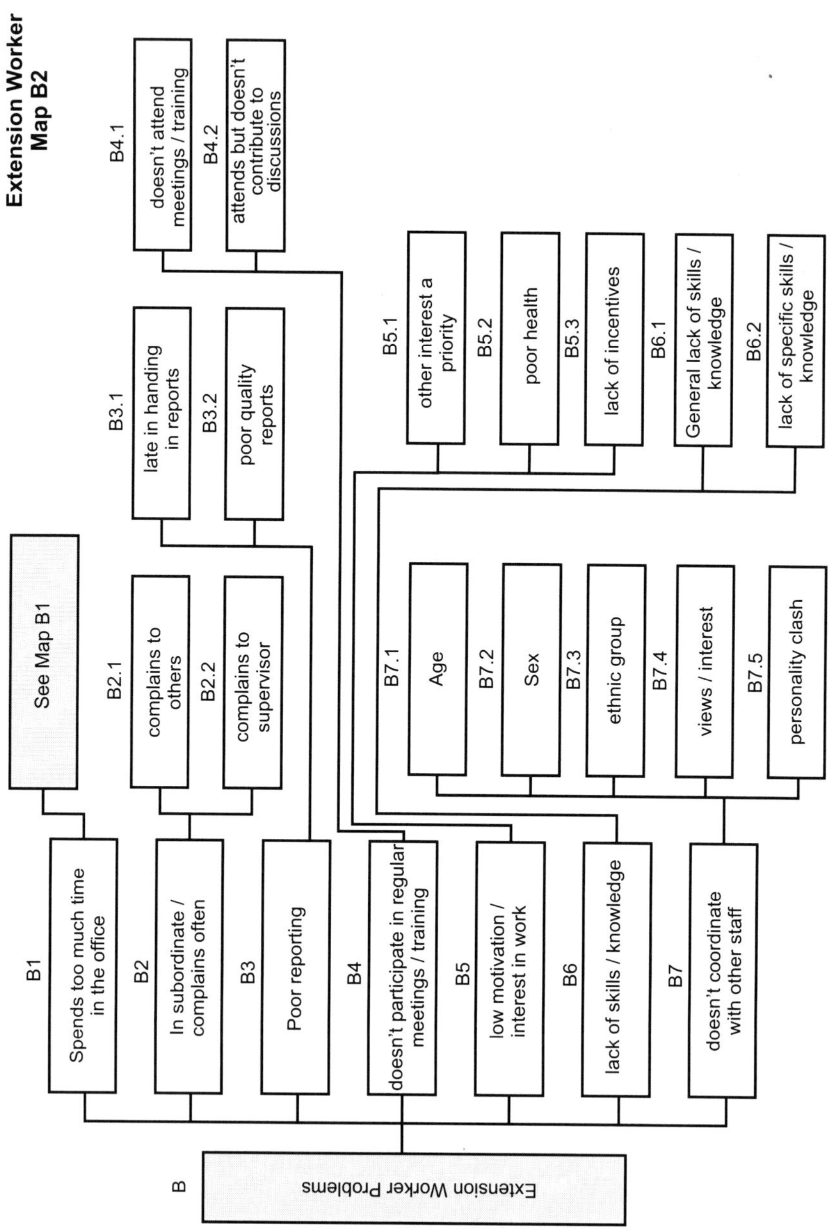
Extension Worker
Map B2
B
Extension Worker Problems
B1
Spends too much time in the office
B1.1
See Map B1
B2
In subordinate / complains often
B2.1
complains to others
B2.2
complains to supervisor
B3
Poor reporting
B3.1
late in handing in reports
B3.2
poor quality reports
B4
doesn't participate in regular meetings / training
B4.1
doesn't attend meetings / training
B4.2
attends but doesn't contribute to discussions
B5
low motivation / interest in work
B5.1
other interest a priority
B5.2
poor health
B5.3
lack of incentives
B6
lack of skills / knowledge
B6.1
General lack of skills / knowledge
B6.2
lack of specific skills / knowledge
B7
doesn't coordinate with other staff
B7.1
Age
B7.2
Sex
B7.3
ethnic group
B7.4
views / interest
B7.5
personality clash

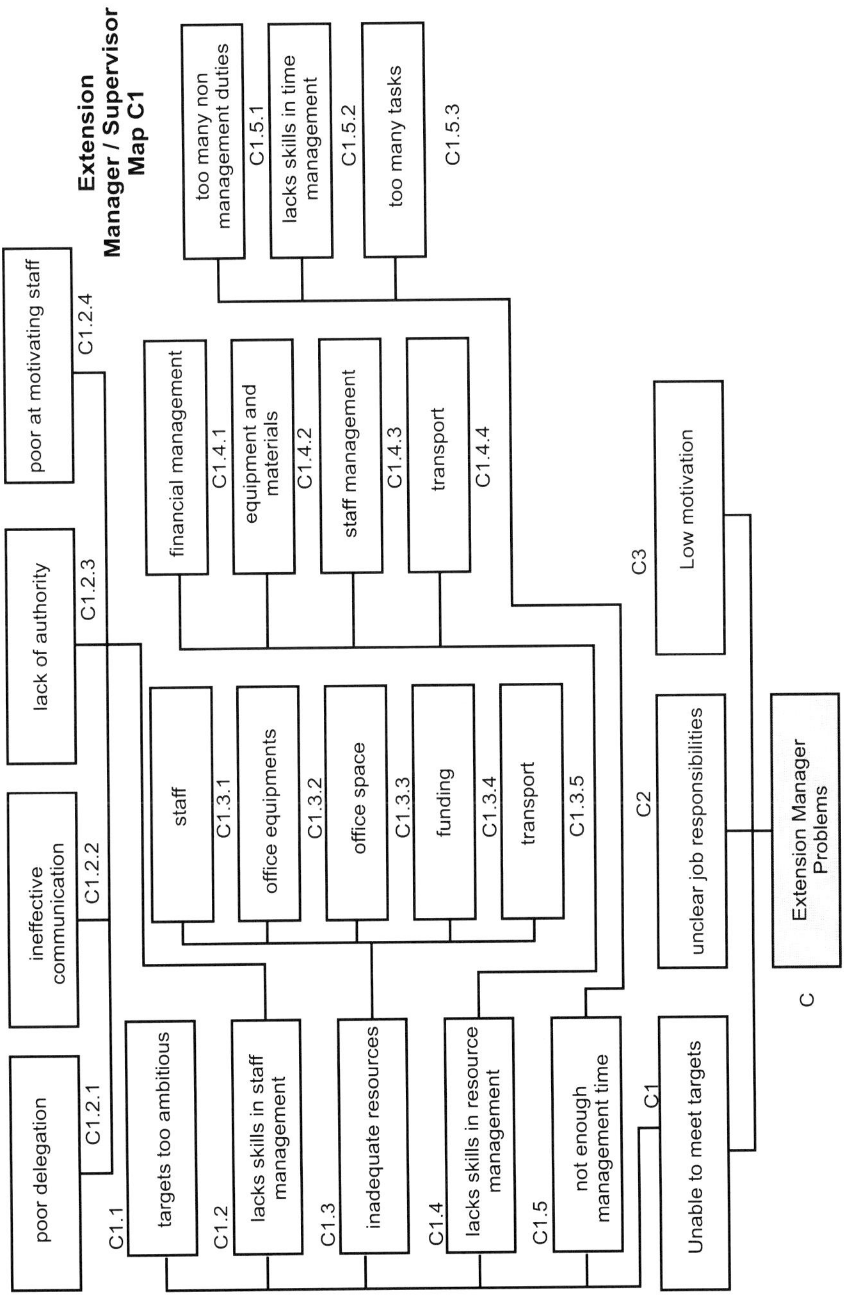
Extension Manager / Supervisor Map C1
Extension Manager Problems
C
Unable to meet targets
C1
unclear job responsibilities
C2
Low motivation
C3
targets too ambitious
C1.1
lacks skills in staff management
C1.2
inadequate resources
C1.3
lacks skills in resource management
C1.4
not enough management time
C1.5
poor delegation
C1.2.1
ineffective communication
C1.2.2
lack of authority
C1.2.3
poor at motivating staff
C1.2.4
staff
C1.3.1
office equipments
C1.3.2
office space
C1.3.3
funding
C1.3.4
transport
C1.3.5
financial management
C1.4.1
equipment and materials
C1.4.2
staff management
C1.4.3
transport
C1.4.4
too many non management duties
C1.5.1
lacks skills in time management
C1.5.2
too many tasks
C1.5.3

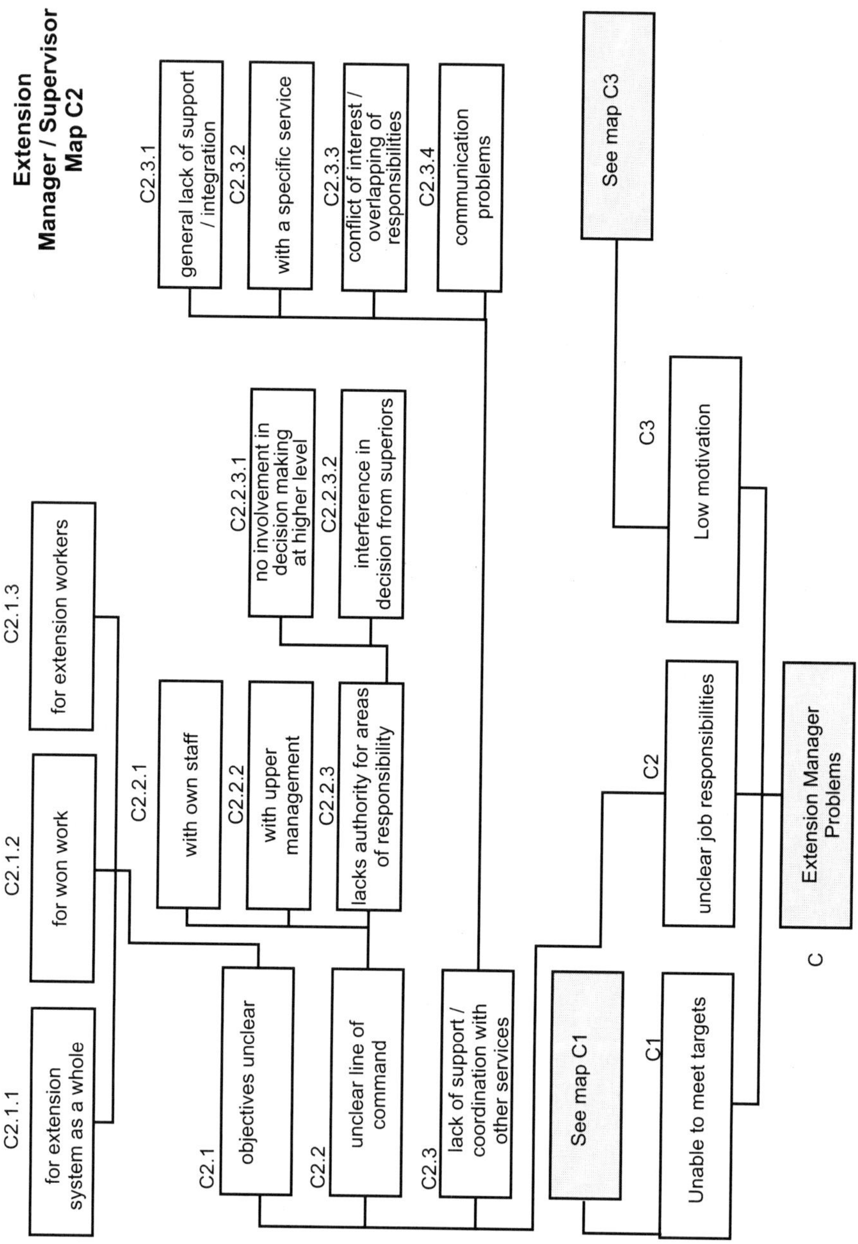
Extension Manager / Supervisor Map C2
C
Extension Manager Problems
C1
Unable to meet targets
See map C1
C2
unclear job responsibilities
C3
Low motivation
See map C3
C2.1
objectives unclear
C2.2
unclear line of command
C2.3
lack of support / coordination with other services
C2.1.1
for extension system as a whole
C2.1.2
for won work
C2.1.3
for extension workers
C2.2.1
with own staff
C2.2.2
with upper management
C2.2.3
lacks authority for areas of responsibility
C2.2.3.1
no involvement in decision making at higher level
C2.2.3.2
interference in decision from superiors
C2.3.1
general lack of support / integration
C2.3.2
with a specific service
C2.3.3
conflict of interest / overlapping of responsibilities
C2.3.4
communication problems

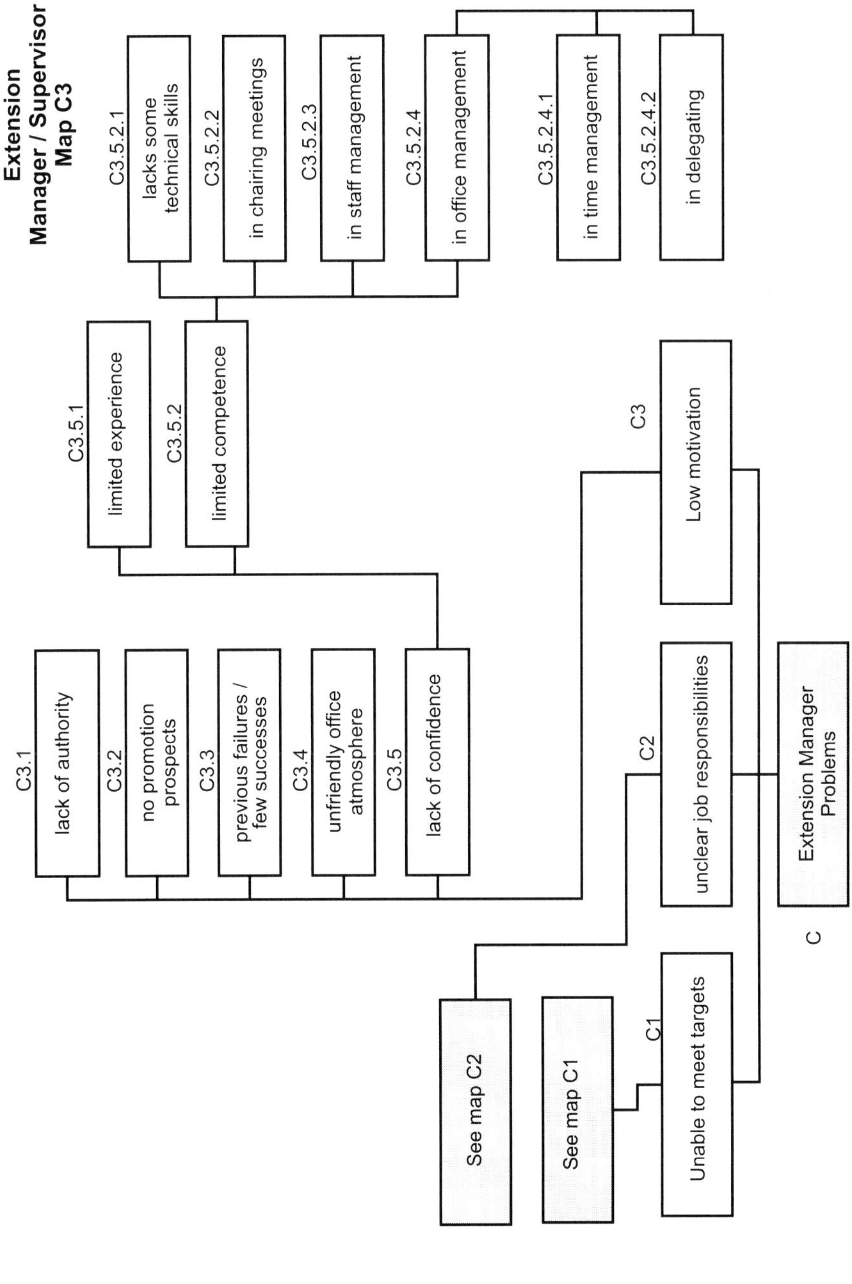
Extension
Manager / Supervisor
Map C3
C
Extension Manager Problems
C1
Unable to meet targets
See map C1
C2
unclear job responsibilities
See map C2
C3
Low motivation
C3.1
lack of authority
C3.2
no promotion prospects
C3.3
previous failures / few successes
C3.4
unfriendly office atmosphere
C3.5
lack of confidence
C3.5.1
limited experience
C3.5.2
limited competence
C3.5.2.1
lacks some technical skills
C3.5.2.2
in chairing meetings
C3.5.2.3
in staff management
C3.5.2.4
in office management
C3.5.2.4.1
in time management
C3.5.2.4.2
in delegating

FARMER CENTRED PROBLEMS

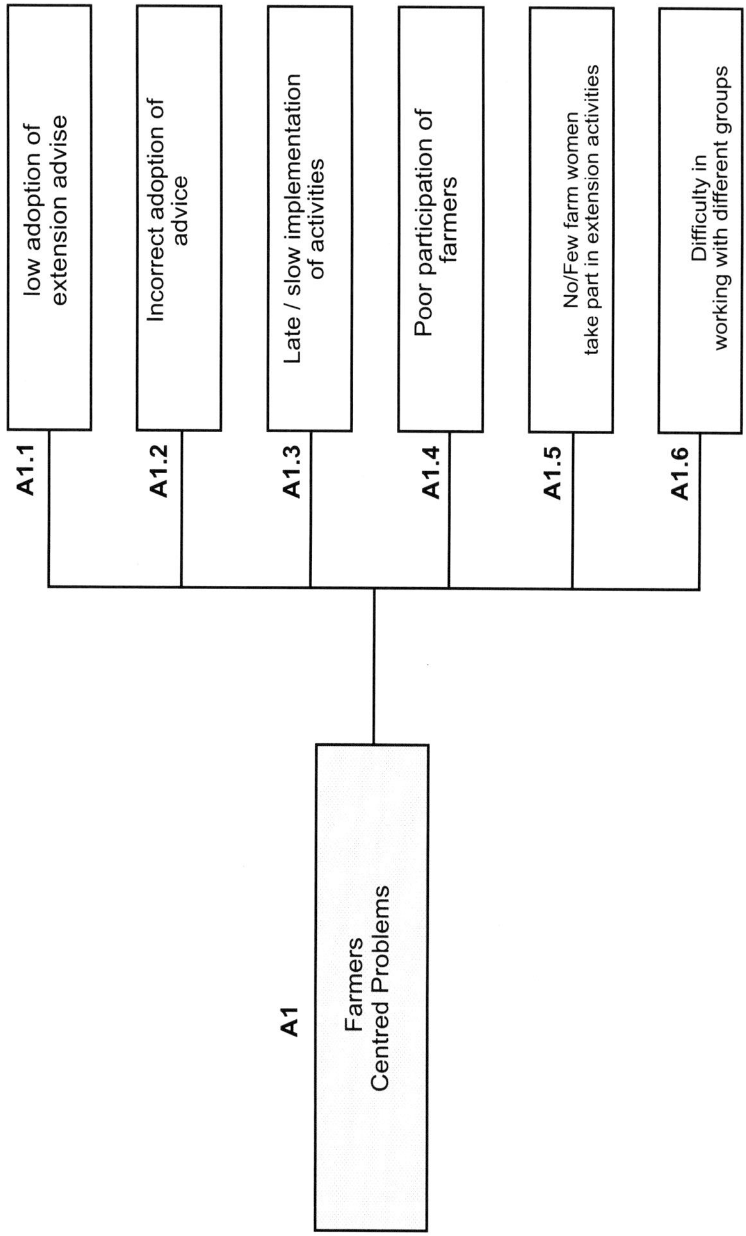

FARMER CENTRED PROBLEMS

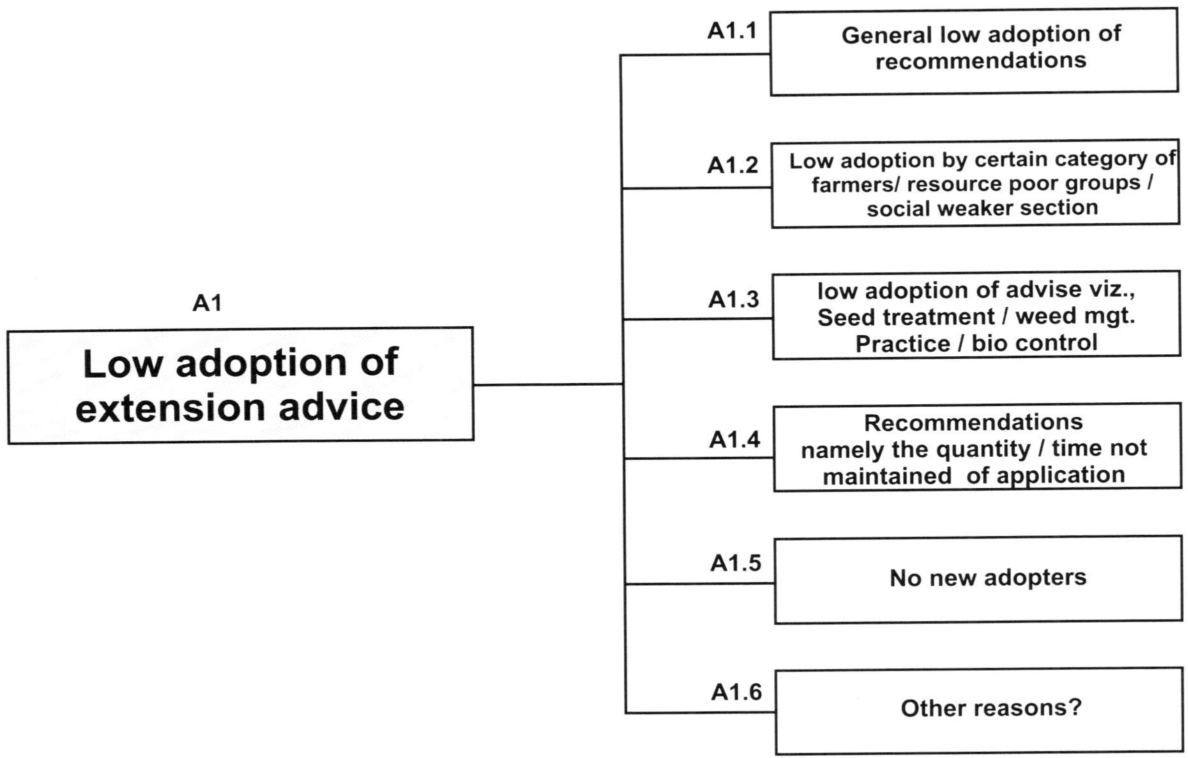

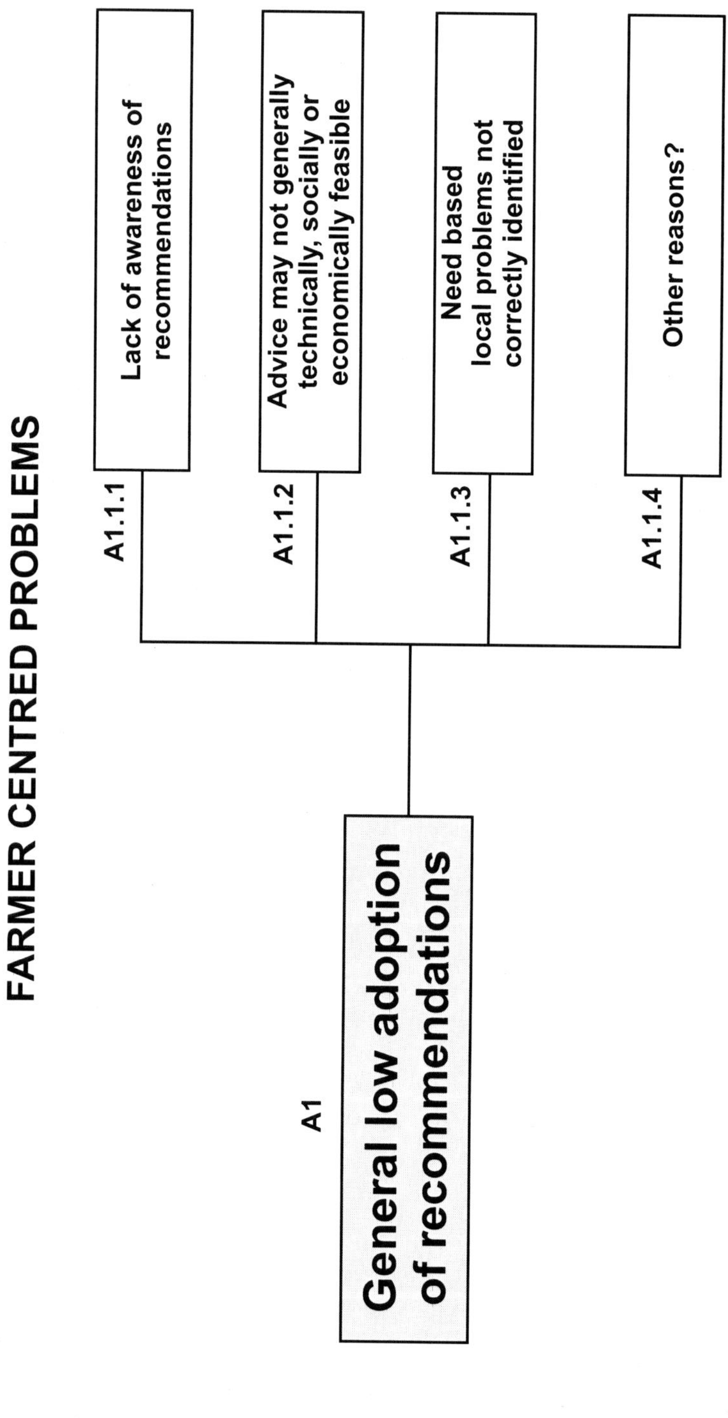
FARMER CENTRED PROBLEMS
A1
General low adoption of recommendations
A1.1.1
Lack of awareness of recommendations
A1.1.2
Advice may not generally technically, socially or economically feasible
A1.1.3
Need based local problems not correctly identified
A1.1.4
Other reasons?

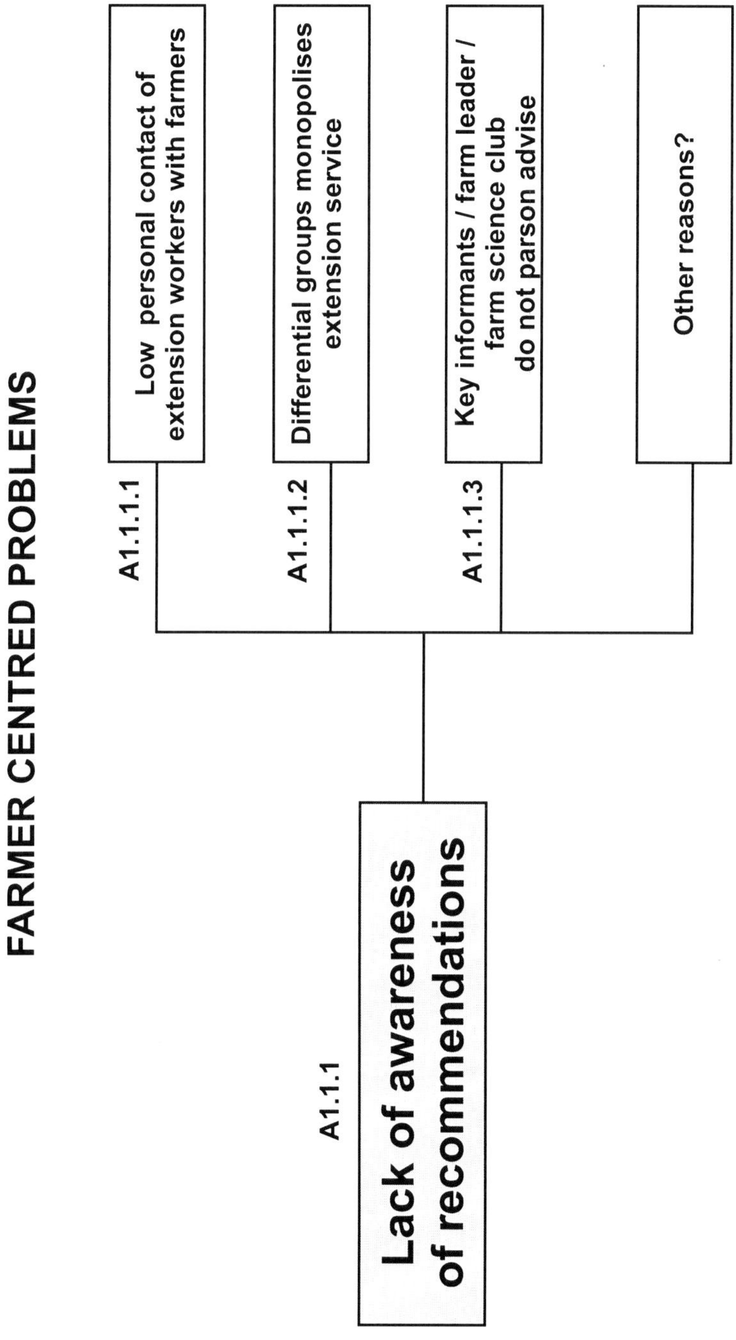
FARMER CENTRED PROBLEMS
A1.1.1
Lack of awareness of recommendations
A1.1.1.1
Low personal contact of extension workers with farmers
A1.1.1.2
Differential groups monopolises extension service
A1.1.1.3
Key informants / farm leader / farm science club do not parson advise
Other reasons?

FARMER CENTRED PROBLEMS

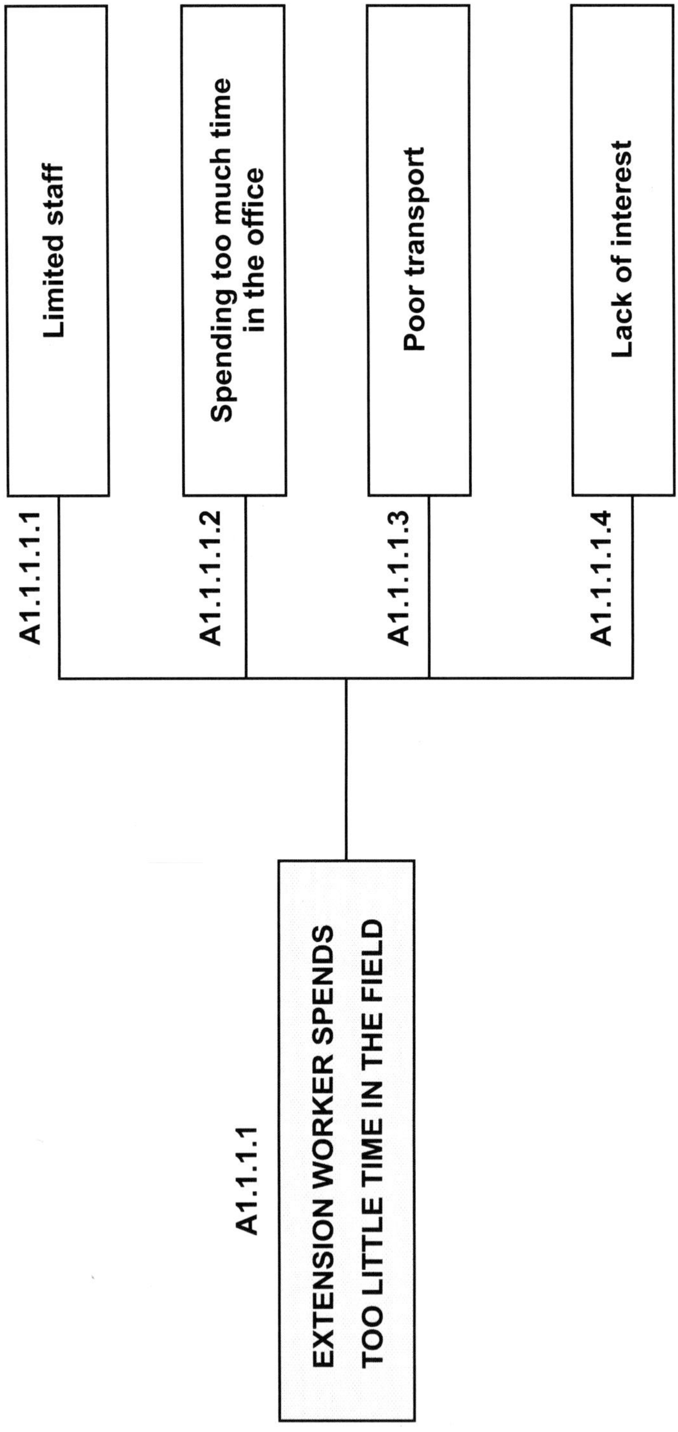

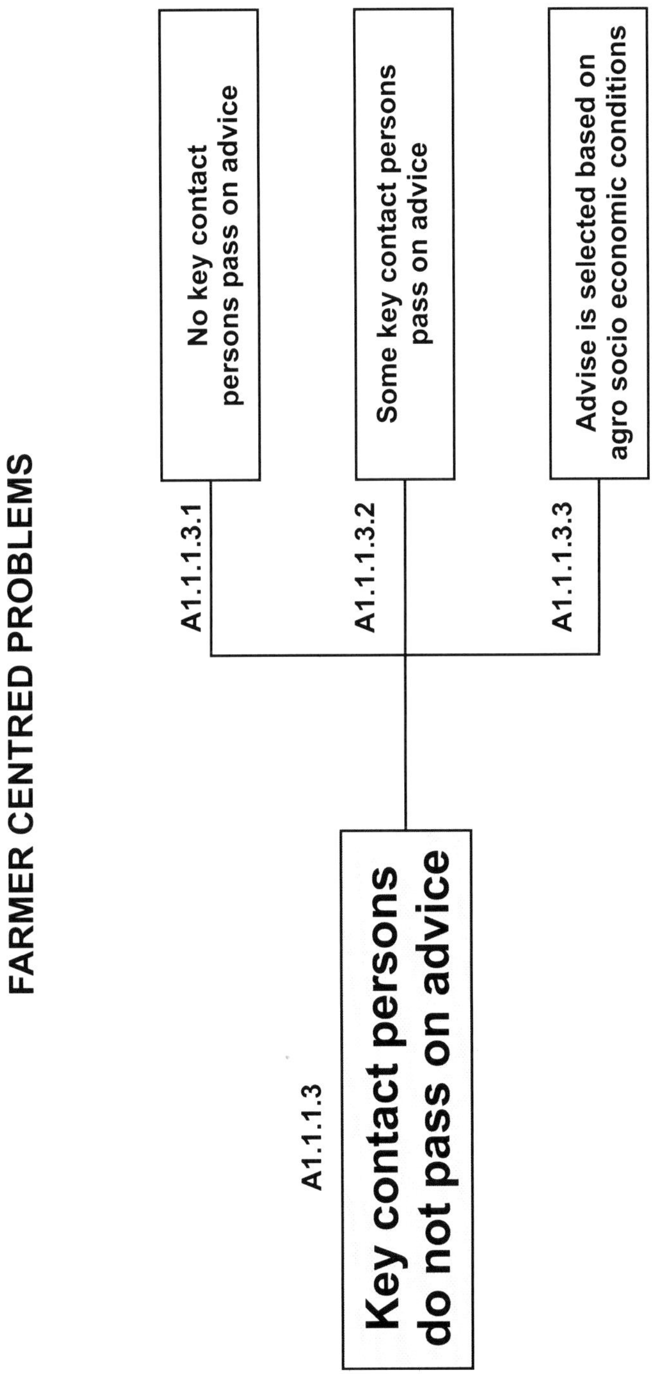
FARMER CENTRED PROBLEMS
A1.1.1.3
Key contact persons do not pass on advice
A1.1.1.3.1
No key contact persons pass on advice
A1.1.1.3.2
Some key contact persons pass on advice
A1.1.1.3.3
Advise is selected based on agro socio economic conditions

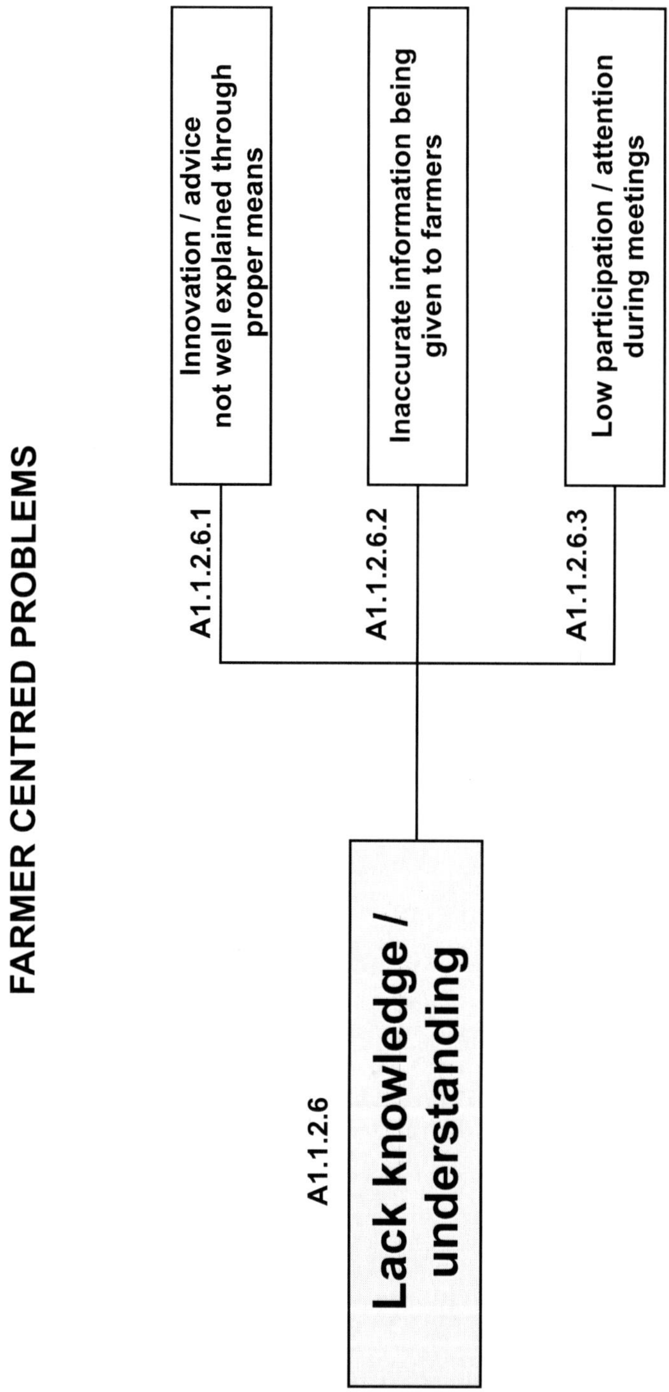
FARMER CENTRED PROBLEMS
A1.1.2.6
Lack knowledge / understanding
A1.1.2.6.1
Innovation / advice not well explained through proper means
A1.1.2.6.2
Inaccurate information being given to farmers
A1.1.2.6.3
Low participation / attention during meetings

Index

Related Books

ISBN: 9789389061536

ISBN: 9789388812870

ISBN: 9789388449908

ISBN: 9789388812900

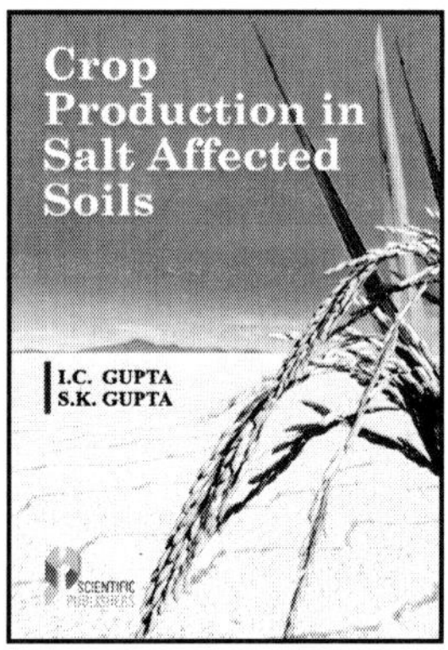

ISBN: 9789387893832

ISBN: 9789388449007

ISBN: 9789388043625

ISBN: 9789387893887

ISBN: 9789388449373